Foundations of Computing

An Accessible Introduction to Formal Languages

Charles D. Allison

Foundations of Computing

An Accessible Introduction to Formal Languages

Charles D. Allison

ISBN 978-0-578-94417-3

Leanpub

This is a Leanpub book. Leanpub empowers authors and publishers with the Lean Publishing process. Lean Publishing is the act of publishing an in-progress ebook using lightweight tools and many iterations to get reader feedback, pivot until you have the right book and build traction once you do.

Cover art, "Feather Moon", by Doug Hamilton, doughamiltonart.com.

To Lexi Lynn,
Thanks for your help,
and for just being you.
Uncle Chuck

To my wife, Sandra, for a lifetime of support and joy.

Contents

Foreword

Computers are everywhere these days—from our phones, to our cars, to appliances in our homes. Indeed, computers, and the programs that run on them, influence every aspect of our lives. It's difficult to imagine what the world was like without them—the entire field of computing is less than one hundred years old. The advances in computing are truly mind-boggling.

Yet none of this would have been possible without a sound theoretical foundation. Alan Turing, Alonzo Church, John von Neumann and others developed the mathematical formalisms that make our modern programming languages possible. All modern programming languages, from popular Python to the venerable C programming language, have their roots therein.

The theory of computation doesn't have to be hard to understand; it has an elegance and simplicity that makes it approachable. In this way, it is much like Einstein's theory of special relativity, embodied by the famous equation, $E = mc^2$. But in both cases, the simplicity of the expression of the theory veils the years of toil required to derive the theory. Fortunately, one can learn the theory without all the hard work of initial derivation. We truly stand on the shoulders of giants.

Many computer science students find this material inscrutable, however. Here is where this book comes in: it presents the theory in a logical, gentle manner, with applications, making it much easier to understand. As I read a draft of the book, I actually became excited about the presentation of the material. At long last there is a book that lays out the theory in a way that it should be. And the results speak for themselves. When Professor Chuck Allison began using drafts of this book in his classes, student performance on assignments and exams went up significantly.

Some students may wonder why they should even bother learning the theory of computation. "After all," they point out, "I'm just going to write application code for websites and mobile apps. I don't need all this theory stuff." That's a fair question. I freely admit that in my twenty years of industry programming, I never wrote a Turing machine. But it was a bit surprising how many times I found myself writing little programming languages and programs to interpret and process them. Foundational computing theory gives you the tools that enable you to do things like that. In addition, most of my industry work was in communication systems, where the software consists of extremely complex finite-state machines, which this book covers in some depth.

As you approach this book, you will notice that it starts very simply. Each concept is liberally illustrated with examples and visualizations. As it progresses, the examples gradually become more complex. When you reach the end, you will likely be surprised at the complexity of the examples you will be able to understand. And that is the essence of learning: through study and practice, you can accomplish things you never thought possible. Enjoy the journey.

– *Neil Harrison*, Computer Science Department Chair, Utah Valley University

Preface

I am personally convinced that any science progresses as much by the writing of better textbooks as by the generation of new knowledge, because good textbooks are what allows the next generation to learn the older stuff quickly and well so we can move on to interesting new things instead of having to painstakingly decipher the discoveries of our forebears.

– Jaap Weel

I wrote this book to make the fundamentals of the theory of computation more accessible to the current generation of undergraduate computer science students. Over the course of twenty years of teaching this subject at Utah Valley University (UVU) in Orem, Utah, I have noticed that computer science undergraduates are not as mathematically sophisticated as in the early days of computer science education, when CS students mostly came from mathematics and the sciences, yet the demand for CS graduates has increased dramatically, with enrollments at UVU and elsewhere following suit. Furthermore, the areas where computing applies to daily life have broadened to the point that society can't seem to do without the convenience that computerized devices afford. CS graduates can make a noticeable contribution to the world of computing work without as much mastery of formal mathematics as was required in times past.

Yet it is crucial that a practicing software engineer understand the nature and limits of computation; the computing practitioner must know what software can and can't do. In this book, I attempt to illuminate the fundamentals of computing for as large an audience as possible. The key results of the theory of computation are covered with numerous examples and sometimes even with working Python programs, while at the same time achieving a measure of rigor with only a minimum of mathematical formalism. Formal proofs appear infrequently, but constructive arguments abound. I use recursive definitions, but use mathematical induction only twice. While mathematical notation appears as needed, I use visual representations profusely to illustrate concepts.

I assume that readers have basic programming knowledge as one would encounter in typical CS1 and CS2 courses, and familiarity with introductory discrete mathematics, including sets, relations, functions, modular arithmetic, the notion of a logarithm, first-order predicate logic, and the structure of simple, directed graphs.

The emphasis is on the structure, properties, and application of formal languages, along with an introductory exposure to computability. Complexity theory is not covered, as it is the topic of a separate undergraduate course on algorithms at UVU. Parsing is only touched on since that topic is covered exhaustively in our compiler course required of CS majors.

I use a uniform approach to transition graphs for the different types of finite-state machines so students feel at home when new features are added. For example, for pushdown automata, I use the same graph notation as for finite automata, only adding stack operations, and only pop one symbol at a time. In addition, when stack start symbols are used, I explicitly push them in the graph, minimizing changes to what students are already accustomed to with finite automata. Furthermore, I require both final/accepting states *and* an empty stack simultaneously for PDA acceptance. This avoids needless discussion of how to convert acceptance by empty stack to and from acceptance by accepting state only, and makes it easier to show that deterministic pushdown automata are closed under complement, as well as making converting PDAs to context-free grammars less exceptional.

Several programming examples appear, and a small number of modest programming projects appear in the exercises. I use Python 3. Python is among the most readable of programming languages; it is often referred to as "executable pseudocode." It is in widespread use in many computing domains and is universally available. An effective innovation is the use of Python programs to teach reductions of one unsolvable problem to another in a compact, inroductory chapter on computability.

I wish to acknowledge the helpful comments of reviewers Matt Bailey, Tyler Barney, Eric Belliston, Kevin Black, Curtis Boland, Alan Chavez, Josh Chevrier, Thomas Christiansen, Cameron Clay, Ethan Clegg, Brennen Davis, Daniel Dayley, Michael Elliott, Ryan Ferguson, Phillip Goettman, Drake Harper, Quinn Heap, Gregory Hodgson, Chad Holt, Preston Hrubes, Joel Johnson, Josh Jones, Ethan Kikuchi, Jared Leishman, Ryan Lever, Caleb MacDonald, Natalie Madsen, Nate Madsen, Wesley Mangrum, Clifton Mayhew, Juan Medrano, Michael Mickelson, James Miles, LuAnne Mitchell, Kelly Nicholes, Eric Nielson, Kris Olson, Timothy Marc Owens, Kailee Parkinson, Cody Powell, Jake Prestwich, Zachary Price, Andrew Ripley, Devin Rowley, Drew Royster, Gail Sanders, Vani Satyanarayan, Blaine Sorenson, Benjamin Thornhill and Jeremy Warren, all UVU alumni. A special thanks to Dr. Keith B. Olson, my long-time colleague and fellow mathematician, for his edits, insights, and encouragement. I also drew inspiration from Dr. Peter J. Downey, who taught me this subject decades ago in CS 473 at the University of Arizona.

I am indebted to Doug Hamilton for the cover art and to Alexa Schulte for assistance in the cover design.

I hope CS students will find this book helpful in understanding what computers can and cannot do, in gaining insights into crafting programs for problems that involve various states or text processing, and in comprehending more fully the nature of computation.

- Chuck Allison, 2016–2021

1. Introduction

There is nothing so practical as a good theory.

– Kurt Lewin

"Charlecote Park Formal Gardens" by Tony Hisgett, licensed under CC BY 2.0

Computer science predates computers. The theoretical foundations of computation were laid long before the first computer was built[1]. What we now call the *theory of computation* was developed in the mid-twentieth century by researchers seeking a systematic method of solving all mathematical problems[2]. Their efforts led to the design of digital computers, and equally importantly, revealed what automatic computation *cannot* do. But first, let's be clear on exactly what we mean by *computation.*

A *computation* is the execution of a *procedure* that processes *input,* unaided by outside intervention once it starts, and, if all goes well, eventually halts, yielding *output* that corresponds to the input given. For example, running a compiler on source code as input and receiving an executable file as output is a computation. A computation that always halts is called an *algorithm.* (We will see computations that do not halt.) The common metaphor for such a procedure is a "black box", as depicted in Figure 1–1.

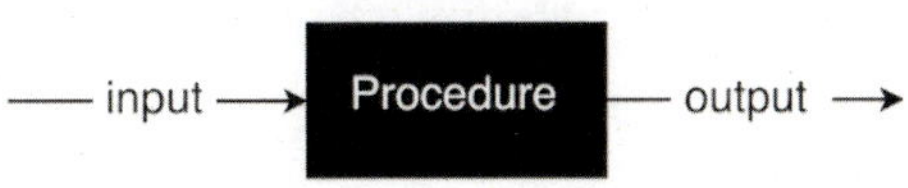

Figure 1–1: Computation as a Black Box

[1]Babbage's Analytical engine of 1837 is considered the first manifestation of a design with the computational power of general-purpose computers. The idea of an algorithm, which is the essence of computer science, is centuries old, and many automated mechanical processes got their start during the first Industrial Revolution. Here we are mainly interested in ideas more abstract, as in software.

[2]It is interesting that their research discovered that no such method exists!

For our purposes, a computation invokes a procedure that represents some underlying mathematical model—a computing mechanism that we call an *abstract machine*[3]. A computation requires a symbolic notation, or *language*, to express it. A language is a *set of strings* with some inherent structure.

In computer programming, we need to agree on which strings of symbols are valid for describing programs and data. Source code is just a string of symbols fed into a compiler and needs to conform to the syntax of the programming language used. Ultimately, input, output, and procedures are expressed by streams of symbols from some *alphabet*. For many programming languages, the alphabet is the set of Unicode characters. For a computer, it is the two-symbol alphabet $\{0, 1\}$. The fact that we think of the symbols 0 and 1 as numbers is immaterial. We are not concerned in this book about the notion of data types; we just process strings, usually one symbol at a time. With this simple approach, one can grasp all that is possible by automatic computation. Types are abstractions built upon this elementary foundation.

1.1 Formal Languages

A *formal language* is a set of strings formed with symbols from some finite alphabet. In studying formal languages, we are interested in the many ways symbols may be validly combined to form strings in given contexts. We talk about *tokens* (or *words*), *sentences* (sequences of words), and language *structure* (how strings may be properly arranged). We do not concern ourselves here with the semantics of the words—only the structure of the strings.

A **formal language** is a *set of strings* formed with symbols from some *finite alphabet.*

To illustrate, consider the simple alphabet[4] $\Sigma = \{a, b\}$. Strings such as *bab* and *abbaab* are among the infinite number of strings that can be formed with these two symbols. We can also choose *none* of the symbols, resulting in the *empty string* (which has length 0), often represented by the string literal "" in programming languages, but which we denote in this book by the lowercase Greek letter, *lambda*: λ. The set of all possible strings using this alphabet is the infinite set depicted below.

$$\{\lambda, a, b, aa, ab, ba, bb, aaa, aab, aba, abb, baa, bab, bba, bbb, aaaa, ...\}$$

Observe the natural way to enumerate this set, which we call **canonical order**. Strings are grouped according to *length*, the groups are arranged in increasing length, and the strings within each group are *alphabetized* (i.e., they appear in lexicographical order based on the *alphabet* used (*a* comes before *b* etc.).

[3] A theoretical model of a computing system. For example, the model for a simple calculator is the basic operations of arithmetic. The abstract machines we cover in this book are various types of finite automata (explained in section 1.2).

[4] It is conventional to use the capital Greek letter Σ (sigma) as an identifier for the current alphabet. We will also use the Greek letters λ (lambda), δ (delta) and Γ (uppercase gamma) in this book, as well as ϕ (phi) for the empty set.

This set is formed by considering all possible ways of concatenating any number of *a*'s and *b*'s together. The set of all possible *concatenations* of elements of an arbitrary set of symbols, *S*, is called the *Kleene star*[5] of that set, and is denoted by S^*. Concatenation of strings is a fundamental operation on formal languages. Since a formal language is a set of strings over some alphabet, a language is therefore a subset of its alphabet's Kleene star.

A formal language over the alphabet Σ is a *subset* of Σ^*.

Suppose the variable x stands for the string bab and y stands for $abbaab$. The table below uses these variables to illustrate common operations on strings in a formal language.

Table 1–1: Fundamental Language Operations

Operation	Name	Example
$\lvert x\rvert$	Length	$\lvert x\rvert = 3$
xy	Concatenation	$xy = bababbaab$
x^n	Repetition	$x^3 = babbabbab$, $x^0 = \lambda$
x^*	Kleene Star	$x^* = \{\lambda, bab, babbab, ...\}$
x^R	Reversal	$y^R = baabba$

Since languages are sets, the usual set operations such as union, intersection, difference, complement, etc., also apply.

Example 1–1

Suppose we want to form $\{bab, abbaab\}^*$, which is the Kleene star of the strings denoted by x and y above. As always, lambda is the first element of a Kleene star. This is followed by all possible concatenations of the original strings, in canonical order:

$$\{\lambda, bab, abbaab, babbab, abbaabbab, bababbaab, babbabbab, ...\}$$

The string *bab* precedes *abbaab* in S^* because of its *length*, not because it appears first in the original set (sets, generally, are unordered, but a canonical ordering is by nature an *ordered set*). We use the *alphabet* to determine the lexicographical *ordering* within each length-group.

Example 1–2

Consider the language, L_e, consisting of all even-length strings over the alphabet $\Sigma = \{a, b\}$:

[5]Usually pronounced "KLAY-nee." Also called "Kleene closure" or "star closure". $aa^* = a^*a$ for any symbol *a*. See https://en.wikipedia.org/wiki/Stephen_Cole_Kleene.

$$L_e = \{\lambda, aa, ab, ba, bb, aaaa, aaab, aaba, aabb, abaa, \cdots\}$$

Another way to describe this language is the Kleene star of the strings from the set $S = \{aa, ab, ba, bb\}$. In other words, $L_e = S^*$. It is obvious that any string in S^* has even length. Take a moment to verify that any even-length string of *a*'s and *b*'s comes from repeated concatenations of elements of *S*.

It is possible to *generate* all the strings in a formal language by using a set of "substitution rules" (aka "rewrite rules") known as a **formal grammar**. A formal grammar consists of substitution rules that stand for substrings[6] that form part of a complete string in a language. A grammar for the language L_e is:

$$E \to aO \mid bO \mid \lambda$$
$$O \to aE \mid bE$$

Since the variable *E* appears first, it is the *start variable.* The vertical bar is an alternation symbol that separates the various ways to substitute for the variable on the left-hand side, so *E* has three substitution rules ($E \to aO, E \to bO, E \to \lambda$) and *O* has two ($O \to aE, O \to bE$). To generate a string, begin by choosing a right-hand side to replace the start variable, *E*, then continue substituting for variables at will, finally ending by choosing a rule containing no variables (in this case, the rule $E \to \lambda$). We can, for example, use this grammar to derive the string *baaa* as follows[7]:

$$E \Rightarrow bO \Rightarrow baE \Rightarrow baaO \Rightarrow baaaE \Rightarrow baaa\lambda \Rightarrow baaa$$

Since lambda is the empty string, it disappears in concatenation. When all we have left are symbols of the alphabet, we are done generating a string.

Another grammar for L_e comes from using the elements of the set *S* from Example 1–1, $\{aa, ab, ba, bb\}$:

$$S \to aaS \mid abS \mid baS \mid bbS \mid \lambda$$

Using this grammar, we can derive the string *baaa* as follows:

$$S \Rightarrow baS \Rightarrow baaaS \Rightarrow baaa$$

---

[6]These intermediate substrings are analogous to phrases of a sentence, so the grammars we study are also known as *phrase-structure grammars.*

[7]In this book we use the single arrow, $\to$, when defining grammar rules, and the double arrow, $\Rightarrow$, when deriving a string using those rules.

Key Terms

computation • abstract machine • algorithm • formal language • alphabet • lambda (λ) • Kleene star • canonical order • formal grammar

1.2 Finite State Machines

We design our computations to only consider one symbol at a time, making decisions based only on the current "state" and the current input symbol. For the language L_e seen earlier, we need only know whether or not the number of symbols consumed at any point is even. Since integers are either even or odd, we have only two states (evenness and oddness) to consider. The **transition graph** in Figure 1–2 depicts a 2-state, abstract machine that will solve our problem; we have only to observe whether or not an input string causes the machine to halt in the "even" state, *E*.

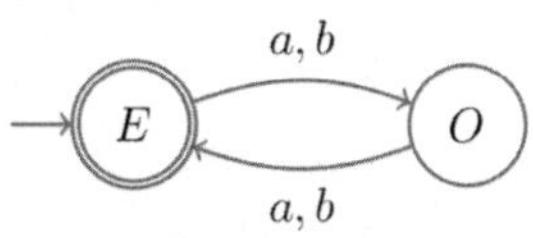

Figure 1–2: A transition graph for a machine that accepts L_e

E is the start state, indicated by an incoming arrow from "out of nowhere." The edges between states *E* and *O* indicate that whenever an *a* or *b* is read, the machine switches state. We combine both *a* and *b* on the same edge instead of having separate edges for each symbol to keep the diagram simple. The double circle for state *E* indicates that it is an *accepting*, or *final* state, meaning that if processing an entire input string ends there, then that string is accepted as part of the language. This machine processes the input string *abba* as follows:

Table 1–2: Processing the string *abba*

State	Remaining Input
E	$abba$
O	bba
E	ba
O	a
E	(accept)

Having terminated in state *E*, the string is accepted. A straightforward Python implementation of this machine follows.

Simulating the machine in Figure 1–2

```
def evenab(s):
    state = 'E'        # Start state
    for c in s:        # Process 1 letter at a time
        match state:
            case 'E':
                state = 'O'
            case 'O':
                state = 'E'
    return 'accepted' if state == 'E' else 'rejected'
```

This machine accepts the empty string because the start state is also an accepting state. This makes sense in this case, since zero is an even number.

> Whether a string is accepted by a finite state machine depends on ending in an accepting state when the last symbol of the string has been read.

Example 1–3

The following machine accepts all and only those strings over the alphabet $\Sigma = \{0, 1\}$ that end with the symbol `1`.

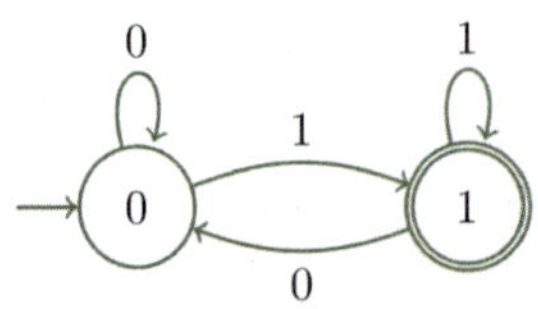

Figure 1–3: Accepts strings ending in 1

Every time a 1 is read, the machine moves to the accepting state. Otherwise it moves to state 0.

Finite state machines, also known as finite **automata** (singular: **automaton**), can give more varied output than just *accept* or *reject*, as illustrated in the next example.

Example 1–4

Removing comments from source code provides an instructive example of an automaton that emits non-trivial output. Suppose a hypothetical language allows comments delimited by a single dollar-sign symbol at each end of a comment's text. An automaton that ignores such comments, but lets uncommented text through as output, appears in Figure 1–4 below.

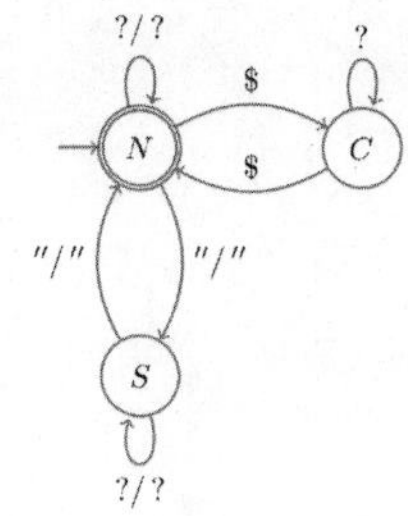

Figure 1-4: State machine to remove $-comments

First, notice that some transition edges have a slash that separates an input symbol from an output symbol; edges without a slash emit no output. (Note: The edges between states N and S have a double-quote symbol as both input and output; the slash is *not* being quoted.) We use a question mark as a special "wildcard" identifier to represent "the current symbol being processed other than the characters explicitly handled by that state"—in this case, other than a dollar sign or double-quote. The start state, N, indicates that we are in "normal" mode, meaning that we are outside of a comment or a string, so we just echo the input symbol to the output stream. When we read a dollar sign outside of a quoted string, we transition to state C, indicating that we are now in a comment. We output nothing until after we return to the start state, having read the comment's closing $. The machine operates similarly for quotes and the state S, except that everything is printed (quotes included). It is not possible to be in comment mode and string mode simultaneously.

Suppose the string `My $fine $fellow, you owe "$1.50"` is fed into this machine. Execution would then proceed as follows (dashes are used to make spaces visible):

State	Remaining Input	Accumulated Output
N	My-$fine-$fellow,-you-owe-"$1.50"	
N	y-$fine-$fellow,-you-owe-"$1.50"	M
N	-$fine-$fellow,-you-owe-"$1.50"	My
N	$fine-$fellow,-you-owe-"$1.50"	My-
C	fine-$fellow,-you-owe-"$1.50"	My-
C	ine-$fellow,-you-owe-"$1.50"	My-
C	ne-$fellow,-you-owe-"$1.50"	My-
C	e-$fellow,-you-owe-"$1.50"	My-
C	-$fellow,-you-owe-"$1.50"	My-
C	$fellow,-you-owe-"$1.50"	My-
N	fellow,-you-owe-"$1.50"	My-
N	ellow,-you-owe-"$1.50"	My-f
N	llow,-you-owe-"$1.50"	My-fe
N	low,-you-owe-"$1.50"	My-fel
N	ow,-you-owe-"$1.50"	My-fell
N	w,-you-owe-"$1.50"	My-fello
N	,-you-owe-"$1.50"	My-fellow
N	-you-owe-"$1.50"	My-fellow,
N	you-owe-"$1.50"	My-fellow,-
N	ou-owe-"$1.50"	My-fellow,-y

State	Remaining Input	Accumulated Output
N	u-owe-“$1.50”	My-fellow,-yo
N	-owe-“$1.50”	My-fellow,-you
N	owe-“$1.50”	My-fellow,-you-
N	we-“$1.50”	My-fellow,-you-o
N	e-“$1.50”	My-fellow,-you-ow
N	-“$1.50”	My-fellow,-you-owe
N	“$1.50”	My-fellow,-you-owe-
S	$1.50”	My-fellow,-you-owe-“
S	1.50”	My-fellow,-you-owe-“$
S	.50”	My-fellow,-you-owe-“$1
S	50”	My-fellow,-you-owe-“$1.
S	0”	My-fellow,-you-owe-“$1.5
S	”	My-fellow,-you-owe-“$1.50
N		My-fellow,-you-owe-“$1.50”

The operation of this machine is also implemented in the following Python code.

Simulates the machine in Figure 1–4

```
def dollar(inputstr):
    output = ""
    state = 'N'
    for c in inputstr:
        match state:
            case 'N':
                if c == '$':
                    state = 'C'
                else:
                    output += c
                    if c == '"':
                        state = 'S'
            case 'C':
                if c == '$':
                    state = 'N'
            case 'S':
                output += c
                if c == '"':
                    state = 'N'
    if state != 'N':
        print('Unbalanced delimiters!')
    return output

print(dollar('My $fine $fellow, you owe "$1.50"'))
```

When executed, this program prints `My fellow, you owe "$1.50"`.

Finite automata apply in many real-life situations. Automatic doors in buildings, vending machines, etc., all have some notion of state and of changing state depending on input stimuli over time.

Example 1–5

The machine in Figure 1–5 below illustrates how to score a tennis game[8]. Each time a player wins a volley, the score progresses through the following point-level sequence: *love* (0), 15, 30, 40, *game.* When the score is tied 40–40, it is called a *deuce.* To win the game from deuce, a player must win two volleys in a row. Winning the first volley from deuce gives the player the "advantage", but that player must win the very next volley to win the game. Otherwise play returns to deuce. This could go on for some time (observe the cycles between the Deuce state and the advantage states (*Ad-*) below).

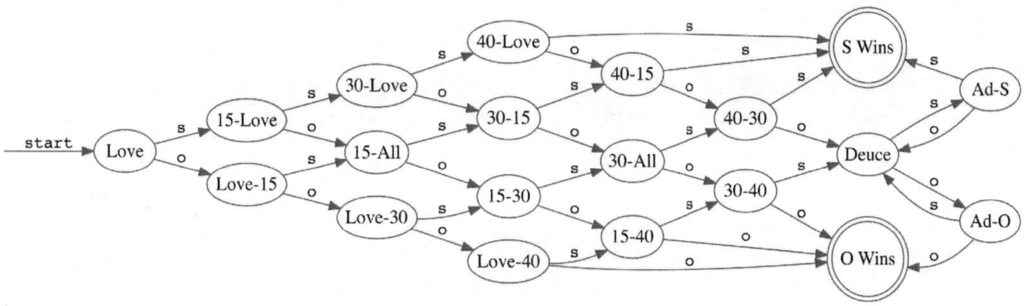

Figure 1–5: Scoring a Game of Tennis

Beginning with the player initially serving, we follow an *s*-edge if the server wins the point or an *o*-edge if the opponent does. There are no out-edges from the states where a player wins, since play ends at that point.

Example 1-6

The diagram in Figure 1-6 models how vending machines from years past (before they had computer chips) kept track of the total amount of money entered. For simplicity, we assume that only nickels, dimes and quarters are accepted and that items cost no more than 50 cents (ignoring any amounts greater than that). The states represent the cumulative total entered at any given point in time.

[8]Diagram based on material from Jeffrey D. Ullman. Used with permission.

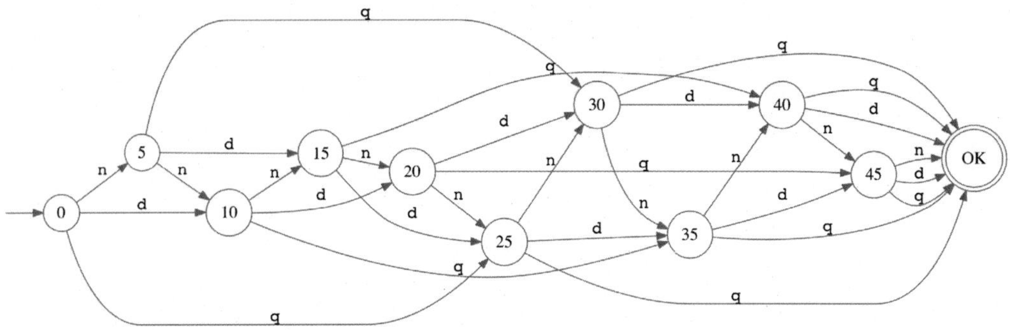

Figure 1–6: Modeling a Vending Machine

Example 1–7

As a final example[9], consider the following schematic of an automatic door.

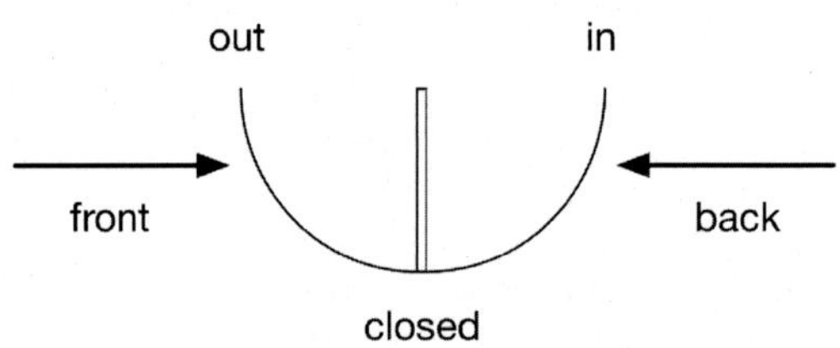

Figure 1–7: A sketch of a 2-way automatic door

This door swings both ways. There are two input stimulus sensors: one when a person walks in from the front (outside), and another when a person exits from the back (inside). If the control mechanism polls for input at small, regular time intervals, there are four possible input configurations over time: **front**, **back**, **both** front and back, and **neither** front nor back. The door can either be opened-*out*, opened-*in*, or *closed*, so these are the three states in our model. The "natural state" is for the door to be closed, so we make that the initial state. We are not testing input strings here, so we have no accepting state. Figure 1-8 below shows an automaton that guarantees that customers can come and go without getting hit by the door. When no one is present within sensor range, the **neither** input signal closes the door.

[9]Example from Sipser, Introduction to the Theory of Computation, 3rd Ed., Cengage, 2013, p. 32.

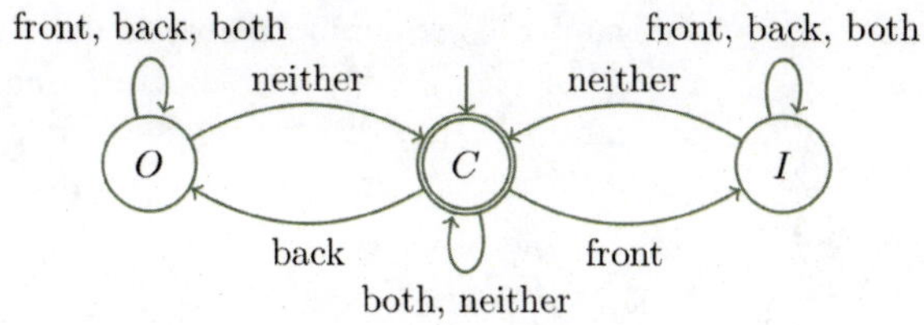

Figure 1–8: Modeling a 2-Way automatic door

Key Terms

state • transition • finite state machine • automaton/automata • transition graph • start/initial state • final/accepting state

---

Exercises

1. List in canonical order the first 11 elements of the subset of $\{0, 1\}^*$ that are of *odd length.*
2. List the first 11 elements of $S = \{aba, ba\}^*$ in canonical order. (Note: the alphabet here is $\{a, b\}$, but the set *S* is being operated upon by the Kleene star.)
3. Draw a finite automaton that accepts all the strings of *odd length* over $\Sigma = \{a, b\}$.
4. Write a formal grammar that generates all strings of *odd length* over $\Sigma = \{a, b\}$.

Chapter Summary

In essence, computation is the meaningful manipulation of symbols. In this book, we relate computations to formal languages, since a computation needs a language to describe its input and output, as well as its operations. Languages can be generated by grammars and recognized by finite-state machines, also called *automata.* The theory of computation comprises three major classes of formal languages along with their associated machines, as shown in the following table.

Language Class	Recognized By	Generated By
Regular	Finite Automata	Regular Grammar Regular Expression
Context-Free	Pushdown Automata	Context-Free Grammar
Recursively Enumerable	Turing Machine	Unrestricted Grammar

As you have seen in this introductory chapter, computation theory is quite practical; it has direct application in digital design, programming languages, compilers—indeed in numerous contexts where an automated process exhibits various states over time. It is no wonder computing has become part of the landscape of our lives. As architect Nicholas Negroponte[10] has said, "Computing is not about computers anymore. It is about living."

Our chief interest is to understand what can be computed in principle, independent of current technology. In Part 3 of this book we will see that the Turing machine is a complete and universal model for automatic computation.

Computation is the meaningful manipulation of *symbols*. Our chief interest is to understand what can be computed *in principle*.

[10]*Being Digital*, 1995.

I Regular Languages

Diagram from csunplugged.com

2. Finite Automata

> *Vernon Conway*: Why is it so difficult for you to accept my orders if you're just a machine?
> *Blue Robot*: Just a machine? That's like saying that you are just an ape.
> – *Automata movie (2014)*

"Vintage Pop Machine Innards" by Richard Eriksson, licensed under CC-BY 2.0

Where Are We?

Language Class	Recognized By	Generated By
Regular	**Finite Automata**	Regular Grammar Regular Expression
Context-Free	Pushdown Automata	Context-Free Grammar
Recursively Enumerable	Turing Machine	Unrestricted Grammar

Chapter Objectives

- *Design finite automata to solve simple computational problems*
- *Convert non-deterministic automata to deterministic ones*
- *Minimize the number of states in a deterministic finite automaton*
- *Use finite automata to model computations that produce output strings*

In this chapter, we look at the class of languages that finite automata accept as well as the computations that finite automata can perform. Finite automata are especially useful in text processing and digital circuit design.

2.1 Deterministic Finite Automata

The automata in Chapter 1 are all *deterministic* in that there is no ambiguity about which edge to follow from each state. Deterministic finite automata are also easy to program; simply have a case for each state, and inside each case, choose the target state corresponding to the current input symbol. In the next section, we will see how non-deterministic finite automata can also be useful.

Definition 2.1

A **deterministic finite automaton** (DFA) is a finite state machine consisting of the following:

- Q, a set of **states**, one of which is the *initial* (or *start*) state, and a subset of which are *final* (or *accepting*) states.
- Σ, an **alphabet** of valid input symbols.
- $\delta : Q \times \Sigma \to Q$, a **function** which, given a state and input symbol, determines the next state of the machine.

Definition 2.1 is merely a mathematical reformulation of what we already understood a DFA to be from Chapter 1. The *transition function*[1], δ, represents the transitions/edges in the machine. The subset of Q comprising the final states can be empty in the case of machines that produce an output string instead of a yes-or-no result. There must be *exactly one* transition out of every state for *each symbol* in the alphabet of the DFA.

There must be *exactly one* transition out of *every* state for *each* symbol in the alphabet of the DFA.

Example 2–1

To illustrate the application of the formal definition of a DFA, consider the DFA in Figure 2–1 below, which accepts all strings over the alphabet $\Sigma = \{a, b\}$ that end with the symbol b.

[1] It is customary to use the Greek letter, δ (delta), since it is commonly used in mathematics to denote change, another name for "transition".

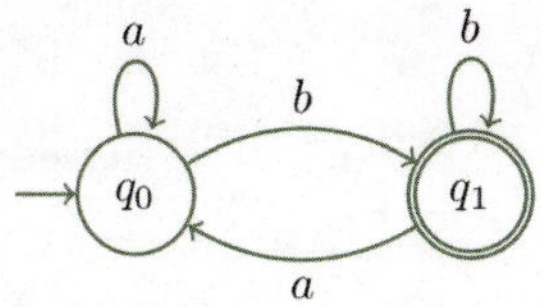

Figure 2–1: DFA accepting strings ending with the symbol *b*

Every time we read a *b*, for all we know it could be the last symbol in the input string, so we move to the accepting state q_1; otherwise we move to state q_0. The formal definition for this DFA is therefore:

- $Q = \{q_0, q_1\}$, where q_0 is the initial state and $\{q_1\}$ is the set of final states
- $\Sigma = \{a, b\}$
- $\delta = \{((q_0, a), q_0), ((q_0, b), q_1), ((q_1, a), q_0), ((q_1, b), q_1)\}$

Yet a third way to depict a DFA is with a *transition table*, as follows:

State	*a*	*b*
q_0	q_0	q_1
$+q_1$	q_0	q_1

In a transition table, the initial state appears first and a plus sign indicates each final state.

Since tables are a common data structure in programming languages, they offer yet another approach to implement DFAs in code. The program below uses Python's dictionary data type to simulate the transitions in the table above. The dictionary named `transitions` maps each `<state,symbol>` input pair to a machine state.

```
def ends_with_b(s):
    transitions = {('q0','a'):'q0', ('q0','b'):'q1',
                   ('q1','a'):'q0', ('q1','b'):'q1'}
    state = 'q0'
    for c in s:
        state = transitions[(state,c)]
    return 'accepted' if state == 'q1' else 'rejected'

print(ends_with_b('aabab')) # accepted
print(ends_with_b('aaba'))  # rejected
```

This program is more concise than using `if` and `else` statements to check states directly in code as we did in Chapter 1.

Example 2–2

Now consider how to construct a DFA that accepts all strings over the alphabet $\Sigma = \{a, b\}$ that end with the substring ab. In cases like this it is often helpful to begin by drawing a "partial machine" that accepts only the required substring:

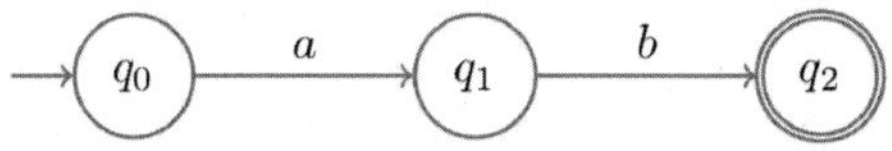

Figure 2–2: Partial DFA accepting strings ending with ab

If we are in state q_0 and read a *b*, we should stay in q_0, since we need an *a* preceding the final *b*. It is only when we read an *a* there that we might have encountered the second-to-the-last symbol in a valid string. State q_1 represents having just read an *a*, so we'll stay there when reading consecutive *a*'s, hence the loops on states q_0 and q_1 in Figure 2–3 below. If we read an *a* in state q_2, it could be the final *a* before the final *b*, so we move back to state q_1. If we encounter a *b* in state q_2, we need to start over by moving to q_0.

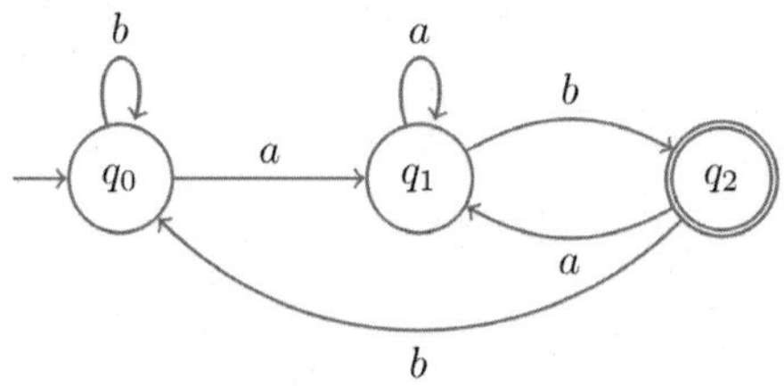

Figure 2–3: Complete DFA accepting strings ending with ab

It is now clear that state q_0 represents having read a b that was *not* preceded by an a.

Example 2–3

To design a DFA that accepts all strings that contain *aba* as a substring, first construct a partial machine that accepts only the string *aba*. (See Figure 2–4 below.)

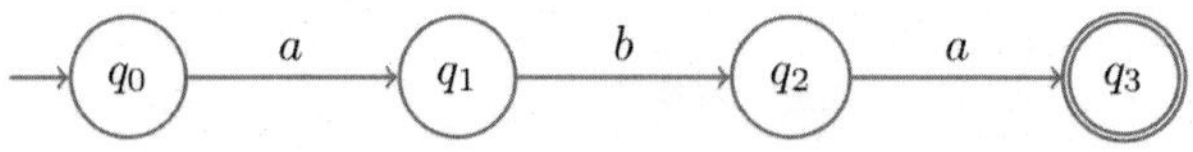

Figure 2–4: Accepts only aba

You can now complete the DFA by filling in the missing edges (see Figure 2–5).

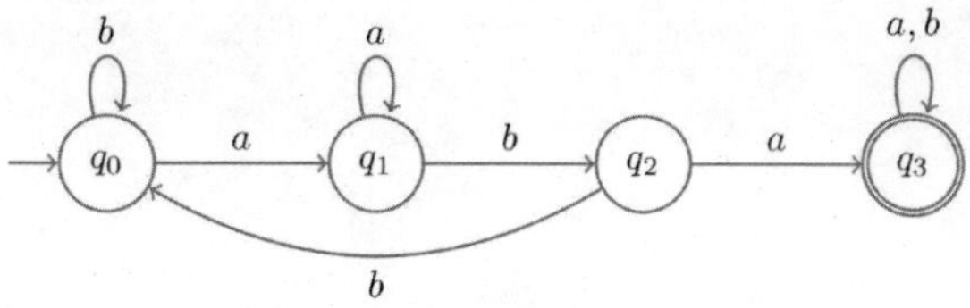

Figure 2–5: Accepts strings containing aba

Example 2–4

Now consider how to define a DFA that accepts the *complement* of the language of Figure 2–5 above (i.e., strings that do *not* contain the substring aba). Once state q_3 is reached, we know that an aba has occurred, so if a string never takes us there, it is in the complement of the language. In general, the complement of language consists of strings that never end in a final state in the original DFA, or, equivalently, that end only in *non-final* states. Therefore, to construct a DFA for the complement of a language, simply **invert the acceptability of each state** in the original DFA. The machine in Figure 2-6 accepts the complement of the language in the previous example.

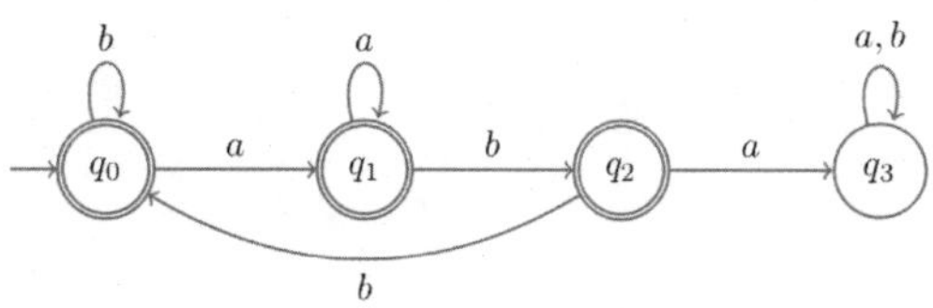

Figure 2–6: Accepts strings not containing aba

There is no going back to an accepting state once a string reaches state q_3. Such a state is called a **dead state** or a **jail state.** Jail states are common in languages that are complements of other languages—in other words, when acceptable strings are defined by properties they *don't* satisfy.

To recognize the *complement* of a language, *invert the acceptability* of each state in a DFA that accepts the original language.

Example 2–5

Another machine that needs a jail state is found in the language of all strings that begin with an a and end with a b. If the first symbol is a b, then there is no hope of the machine accepting it. (See Figure 2–7).

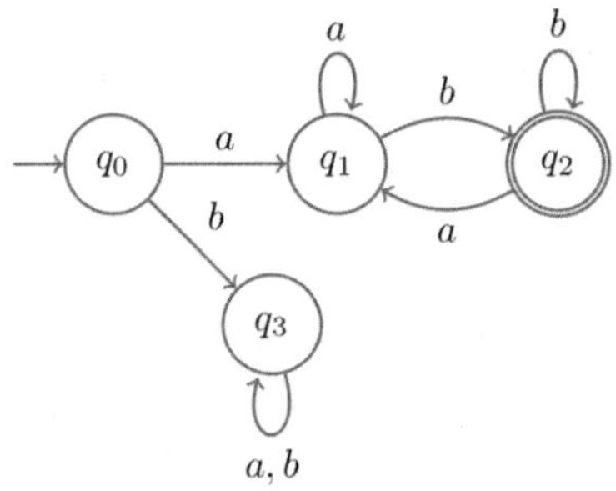

Figure 2–7: A DFA accepting strings starting with a and ending with b

---

Example 2-6

How would we construct a DFA that accepts strings of zeroes and ones where the next-to-last symbol is a 1? Such a string must end with either 10 or 11. Since we never know when the next-to-last input occurs, we need to track those two substrings whenever they occur. See Figure 2–8.

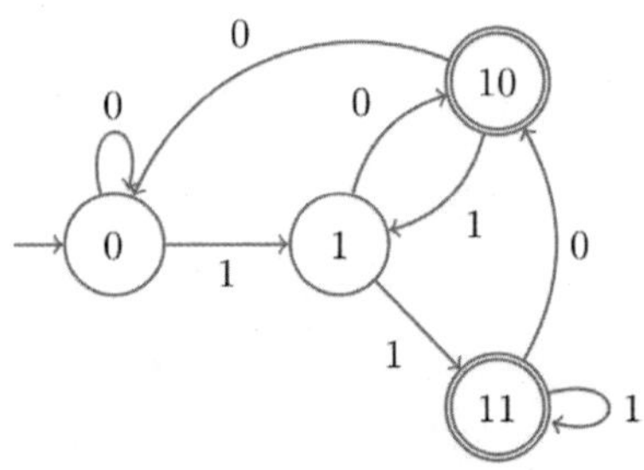

Figure 2–8: Accepts strings with a 1 in the next-to-last position

The state labels reflect the substring just processed that led to each state. Observe how the out-edges from the accepting states behave. The last symbol read reaching those states becomes the next-to-last symbol on the subsequent move.

Example 2-7

Now consider how to recognize strings over $\{0, 1\}$ that end with a 1 in the *third-to-last* position. Reading a 1 starts us on a potential accepting path. From there we will read a 0 or 1, at which point we will have read one of 10 or 11. An acceptable string must be of at least length 3 and end with 100, 101, 110, or 111.

These possibilities are represented by accepting states in the following machine:

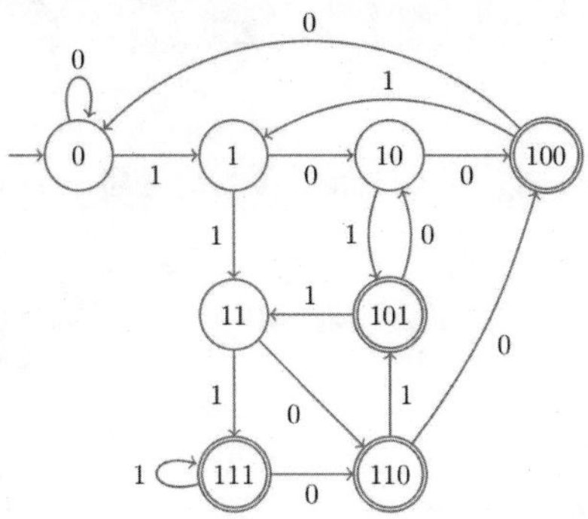

Figure 2–9: A DFA that accepts strings with a 1 in the third-to-last position

Where this machine moves from its accepting states depends on the four possibilities of where the 1's occur in the last two symbols read, which this DFA tracks nicely.

Example 2–8

Consider a machine that reads the bits of a binary number, one at a time, left-to-right, to determine what the modulus (i.e., remainder) of the input number is when divided by 3. We have neither arithmetic instructions nor auxiliary memory—all we have are states—so we must find a way to keep track of the modulus as we go. We need three states, therefore, corresponding to the remainders 0, 1, and 2, with 0 as the initial state (i.e., the number is considered to be zero before any input is consumed). See Figure 2–10.

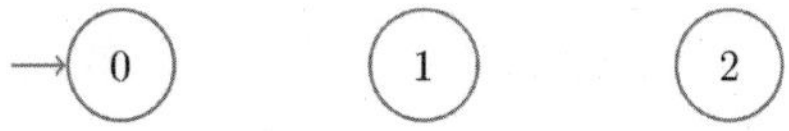

Figure 2–10: States representing possible remainders mod 3

How do we track the remainder as we read a single bit-symbol at a time? In the simplest instance, we can assume that our current number is 0 to start with. Appending a 0 still gives us $00 = 0$, so $\delta(0,0) = 0$. Appending a 1 gives the number $01 = 1$, so $\delta(0,1) = 1$. If instead the current number is 1, the machine is in state-1 (since $1 \equiv 1 \mod 3$), then appending a 0 transforms the number into $10_2 = 2 \equiv 2 \mod 3$, so $\delta(1,0) = 2$. Similarly, appending a 1 gives $11_2 = 3 \equiv 0 \mod 3$, so $\delta(1,1) = 0$. If the current number is $2 = 10_2$, appending a 0 yields $100_2 = 4 \equiv 1 \mod 3$, while appending a 1 results in $101_2 = 5 \equiv 2 \mod 3$, so $\delta(2,0) = 1$ and $\delta(2,1) = 2$. The following DFA records all of these results.

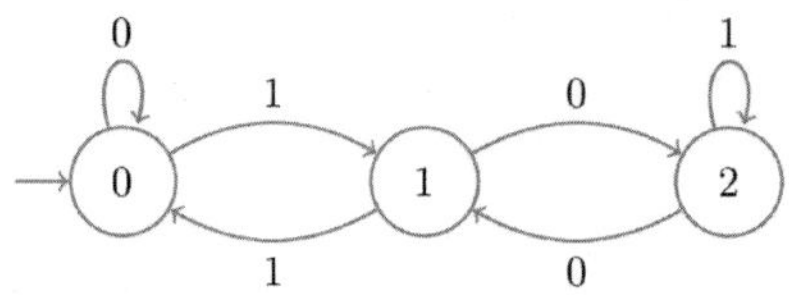

Figure 2–11: DFA whose states identify the residue mod 3 of a binary number

Having established how the machine behaves for the numbers 0, 1, and 2, we now use this information as a base case to show by induction that the machine is correct for any number. The induction hypothesis is that when we are in state-0 that the number is congruent to 0 mod 3, and analogously for states 1 and 2.

To understand how the value of the input number changes as we append the next 0 or 1, consider how to recognize the number 5, say, which appears as 101 in binary. Reading the bits left-to-right, we construct this number a symbol at a time as follows:

1. Start with our number, n, say, initialized to 0, and then read the first 1-bit, "appending" it to our current n. Appending a 1 to a bit string is numerically equivalent to shifting the bits one position to the left (which is the same as multiplying the number by 2), and then adding the 1, giving the number $2 \cdot 0 + 1 = 1$.
2. Next, we read and append the 0-bit, so we shift left (i.e., multiply the running result by 2) and add zero, giving $2 \cdot 1 + 0 = 2 = 10_2$.
3. Finally, we append the last 1-bit to obtain $2 \cdot 2 + 1 = 5 = 101_2$.

To summarize: every time we append a bit, b, we update our number, n, to be $2n + b$, where b is 0 or 1.

We are now able to track the remainders for any bit string. Consider what happens when we are currently in state-0. By the inductive hypothesis, we know that in this state the number must be a multiple of 3, because its remainder is 0, therefore, $n = 3k$ for some k. If we append a 0-bit, then the updated number is $n = 2(3k) + 0 = 6k \equiv 0 \mod 3$, meaning that the number is still a multiple of 3, so we remain in state-0. If instead we append a 1-bit, the number becomes $2(3k) + 1 \equiv 1 \mod 3$, so we move to state-1, as shown in Figure 2–12.

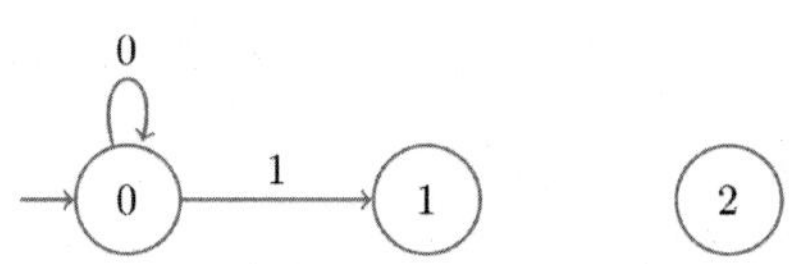

Figure 2–12: Adding edges for state-0

If we are in state-1, then $n \equiv 1 \mod 3 \Rightarrow n = 3k + 1$, for some k, so the results for appending a 0 are $2(3k + 1) + 0 = 6k + 2 = 3(2k) + 2 \equiv 2 \mod 3$, and when appending a 1 we get $2(3k+1)+1 = 6k+3 = 3(2k+1) \equiv 0 \mod 3$. Figure 2–13 updates our working DFA accordingly.

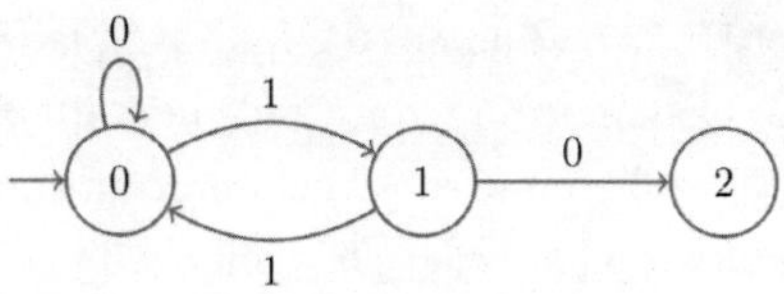

Figure 2–13: Adding edges for state-1

In state-2 we know that $n = 3k+2$, so the results become $2(3k+2)+0 = 6k+4 = 3(2k+1)+1 \equiv 1 \mod 3$ for appending a 0, and appending a 1 gives $2(3k+2)+1 = 6k+5 = 3(2k+1)+2 \equiv 2 \mod 3$. This is consistent with the DFA we first saw in Figure 2–11.

All the languages we have seen so far are examples of what we call *regular languages*, which are simply those languages that are accepted by some DFA.

Definition 2.2

A **regular language** is a formal language for which there exists a deterministic finite automaton that accepts all and only those strings in the language.

Key Terms

determinism • DFA • transition function • jail state • complement • regular language

Exercises

1. Draw a transition graph for the following DFA.

	a	b
q_0	q_0	q_1
q_1	q_0	q_2
$+q_2$	q_1	q_3
q_3	q_2	q_4
q_4	q_3	q_4

2. Draw transition graphs for DFAs that accept the following languages:
   a. Strings over $\Sigma = \{0, 1\}$ that begin with 1 and end with 0.
   b. Strings over $\Sigma = \{a, b\}$ where the substring aa occurs at least twice.
   c. Strings over $\Sigma = \{a, b, c\}$ where every b is followed immediately by at least one c. Any string of a's and c's is accepted if there are no b's at all in the string.
3. Describe in English the language accepted by the following DFA.

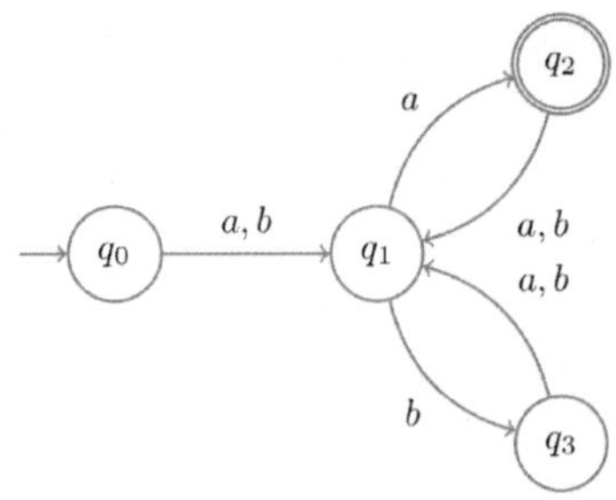

4. The following "incomplete" DFA accepts all strings matching the pattern a^*baba^*, but there are edges missing. Complete the DFA by adding a jail state and the missing edges.

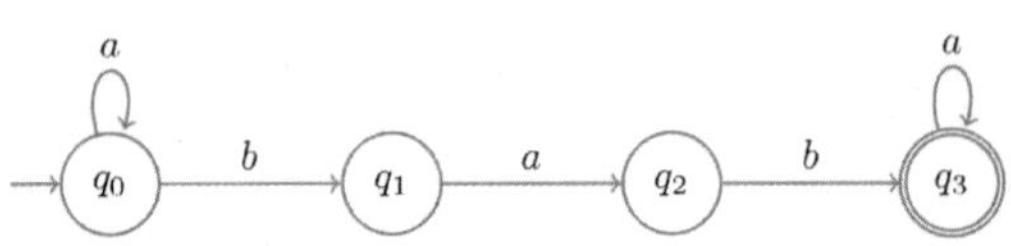

5. Draw a DFA for the language of all strings over $\Sigma = \{a, b\}$ that do *not* have two consecutive b's.
6. Define the transition function, δ, for the DFA in the previous problem.
7. Draw a DFA for the language of all strings over $\Sigma = \{a, b\}$ where no two adjacent symbols are the same. For example, bab and $ababa$ are in the language, but aa and abb are not.
8. Draw a DFA for the language of all strings over $\Sigma = \{a, b\}$ that have an even number of a's and an even number of b's. (*Hint*: You only need four states.)
9. Following Example 2–8, draw a DFA that accepts all bit strings that represent numbers congruent to 3 mod 4.

Programming Exercise

1. Implement the DFA in Figure 2–9. Be sure to reflect states and transitions in your code. Test it with two strings—one that is accepted and one that is rejected.

2.2 Non-Deterministic Finite Automata

The DFA in Figure 2–9 took a lot of effort to design. It is possible, however, to more succinctly express the idea of "ends with a 1 in the third-to-last position" than a DFA allows. With non-determinism, we can get right to the point of expressing exactly what we want, as shown in the *non-deterministic* finite automaton (NFA) in Figure 2–14 below.

Example 2–9

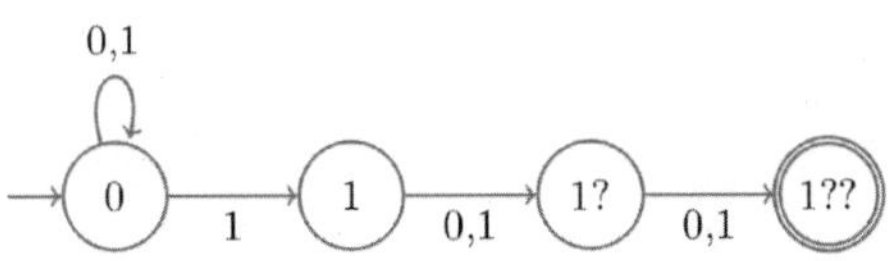

Figure 2–14: An NFA that accepts the language of Figure 2–9

Two things stand out in the transition graph above. First, there are *two choices* of transitions from the initial state for an input of the symbol 1; we may either stay in state-0 or move to state-1. This is what makes this machine non-deterministic. Second, there are *no out-edges* exiting the accepting state. This is because we want to end there. For a string to end in the accepting state, we must choose the edge from state-0 to state-1 at just the right moment. If such a choice can be made, we say that the NFA accepts the string. An accepting path for the input string 10110 is the sequence of states $0, 0, 0, 1, 1?, 1??$. Clearly, only strings with a 1 in the correct position can end in final state. The important thing here is how much easier it is to conceive and express an NFA for this language than a DFA.

This may seem a strange approach to designing finite automata, and programming such a machine may appear to require some sort of backtracking procedure, in case we don't make the right choice (because we don't know ahead of time when an input symbol is the last one in the string). Alas, we can't "back up" in a finite automaton; all decisions must be made based only on the current state and input symbol. Fortunately, it is possible to convert any NFA to a DFA by keeping track of all possible moves as we go. We will show this procedure shortly, but first, let's see some more NFAs.

Example 2–10

The NFA in Figure 2–15 accepts strings of length two or greater that begin and end with the same symbol.

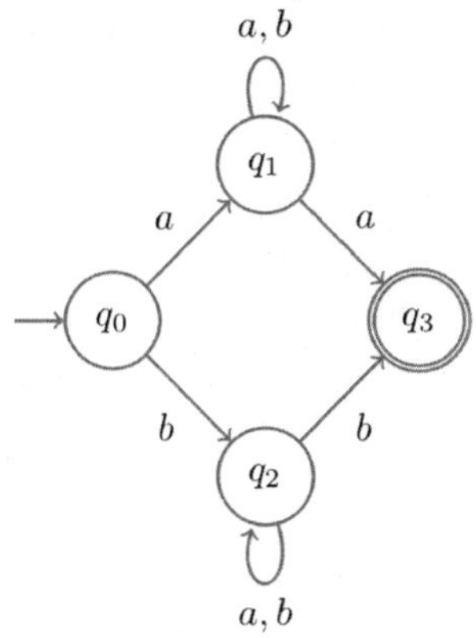

Figure 2–15: NFA accepting strings that begin and end with the same symbol

The non-determinism occurs on the middle states where there are two choices for a and b, respectively. The path to accept the string $bbab$ is q_0, q_2, q_2, q_2, q_3.

It is often easier to use a NFA rather than a DFA to design a finite automaton for a regular language.

Example 2–11

The NFA in Figure 2–16 accepts strings that either contain the substring aa or the substring bb:

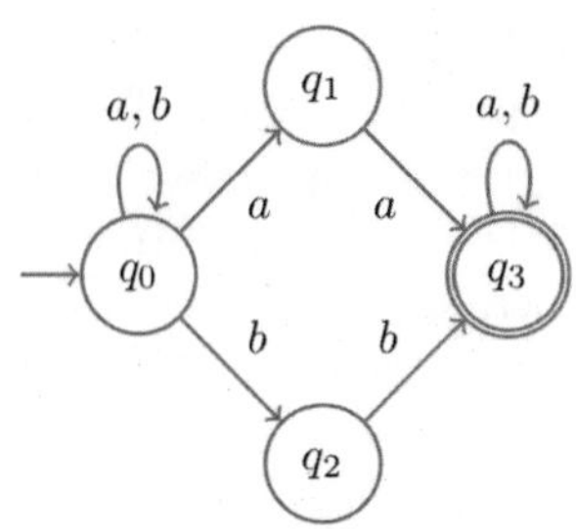

Figure 2–16: NFA that accepts strings containing *aa* or *bb*

It doesn't matter if the string contains both aa and bb. Either path will do in that case. There are multiple accepting paths for the string $baaababba$ for example:

$$q_0, q_0, q_0, q_1, q_3, q_3, q_3, q_3, q_3, q_3$$
$$q_0, q_0, q_1, q_3, q_3, q_3, q_3, q_3, q_3, q_3$$
$$q_0, q_0, q_0, q_0, q_0, q_0, q_2, q_3, q_3, q_3$$

to name only three. (The initial q_0 is where we start before reading input.)

Example 2–12

NFAs have one more interesting feature: they can move from one state to another "on a whim"—in other words, without consuming any input at all. We indicate this by a *lambda transition.* See if you can guess what language the NFA in Figure 2–17 accepts before reading further.

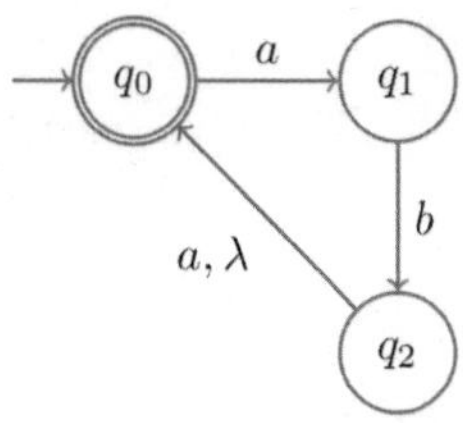

Figure 2–17: What language does this NFA accept?

A lambda transition is basically a "free ride"; we can take it whenever we enter its "from-state" without consuming any further input. That means that for the NFA above, we have a cycle starting and ending with state q_0, where on each round trip we must encounter the substring ab, *optionally* followed by another a. Strings accepted by this NFA include ab,aba,$ababa$,$abababa$, and $abababaab$. We can also represent this language by the expression $\{ab, aba\}^*$, using the Kleene star operator introduced in Chapter 1. In other words, we choose either ab or aba at will until we decide to stop, or we choose nothing at all and obtain the empty string. Both $ababa$ and $abaaba$ can be accepted by the path $q_0, q_1, q_2, q_0, q_1, q_2$.

Example 2–13

Now suppose our alphabet is $\Sigma = \{a, b, c\}$ and we want to accept strings that match the pattern $a^*ac^*b^*$. That is, we need one or more a's followed by zero or more c's followed by zero or more b's. The NFA in Figure 2–18 recognizes this language.

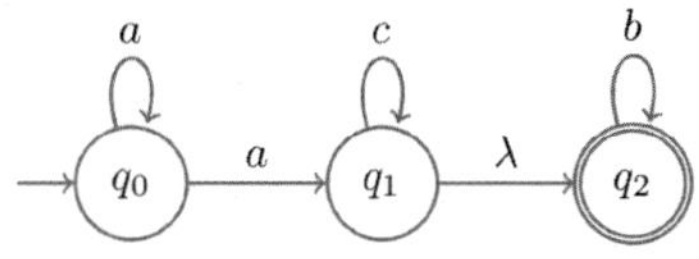

Figure 2–18: NFA accepting $a^*ac^*b^*$

Observe how we represent a Kleene star of a single symbol as a *self-loop* on a state. Also, we need a single edge that is *not* a loop for the required a, because self-loops on states are *optional* moves. The lambda transition above forces the b's to follow the c's while keeping both optional.

We are now able to give a formal definition of a non-deterministic finite automaton.

Definition 2.3

A **non-deterministic finite automaton** (NFA) is a finite state machine consisting of the following:

- Q, a set of **states**, one of which is the initial (or start) state, and a subset of which are final (or accepting) states.
- Σ, an **alphabet** of valid input symbols.
- $\delta : Q \times (\Sigma \cup \lambda) \rightarrow 2^Q$, a **function** which, given a state and an input symbol (which could be λ), determines the next possible **set of states** of the machine to move to (i.e., a *choice* of states).

The differences between DFAs and NFAs occur in the definition of δ, namely:

1. λ is a valid input (i.e., we can move without consuming any input).
2. More than one state may be reached by the same input symbol.
3. It is possible that no transitions exist at all for a particular input (an *implicit jail*).

The notation 2^Q, sometimes written $\mathcal{P}(Q)$, represents the **powerset**[2] of the set of states, Q. The key feature of NFAs is that there may be zero or more out-edges for any input symbol exiting any state, hence the output is a subset of Q. Also, a DFA is just a special case of a NFA, since machines are *not required* to have any lambda transitions or multiple out-edges from any state for the same symbol. NFAs just give us more convenience in designing automata for regular languages.

[2]The powerset of a set is the set of all possible subsets of that set. If $|Q|$ represents the cardinality of the set Q, then the cardinality of the power set is $2^{|Q|}$, hence the exponential notation for powersets. For example, $Q = \{q_0, q_1\} \Rightarrow 2^Q = \{\phi, \{q_0\}, \{q_1\}, \{q_0, q_1\}\}$, and $|2^Q| = 2^{|Q|} = 2^2 = 4$.

When moving from one state to another in a NFA, remember to take into account **lambda transitions** emanating from the newly entered state.

Equivalence of NFAs and DFAs

We now show how to convert the NFA in Figure 2–15 to a DFA. The key idea is that for each state we have a (possibly empty) set of states to track for every input symbol. The tree in Figure 2–19 illustrates how things can progress as we begin with the start state of the original NFA, q_0, and track all possible moves while reading the first three symbols of any input string. (Going three levels deep suffices in this case to observe all the possible states for this language.) Multiple transitions leaving a state with the same symbol appear in horizontally adjacent boxes below.

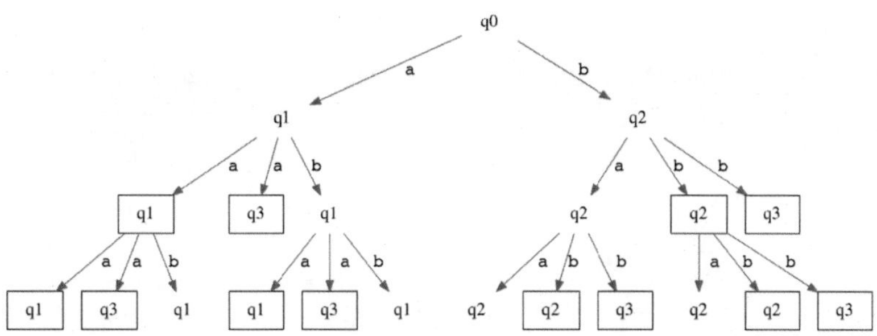

Figure 2–19: Tracking multiple out-edges simultaneously

The symbol a can reach either q_1 or q_3 from state q_1. We will combine these two states into a single, "composite" state, $\{q_1, q_3\}$. The same logic applies to states q_2 and q_3 when leaving state q_2 with a b, giving us the composite state $\{q_2, q_3\}$. See Figure 2–20.

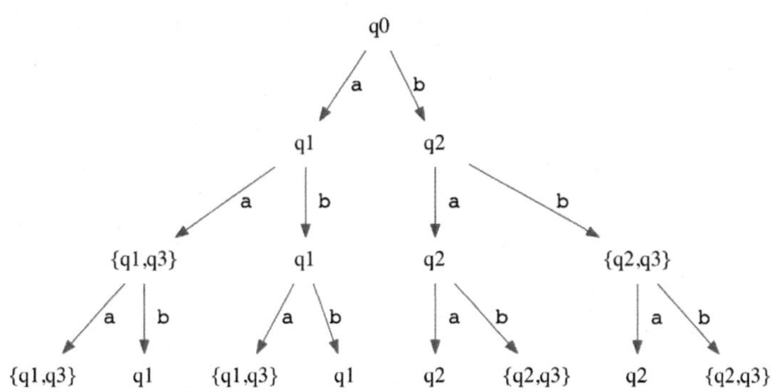

Figure 2–20: Combining like transitions into composite states

This tree suggests that five possible states suffice to process all possible movements: $q_0, q_1, q_2, \{q_1, q_3\}$, and $\{q_2, q_3\}$, giving the following equivalent DFA.

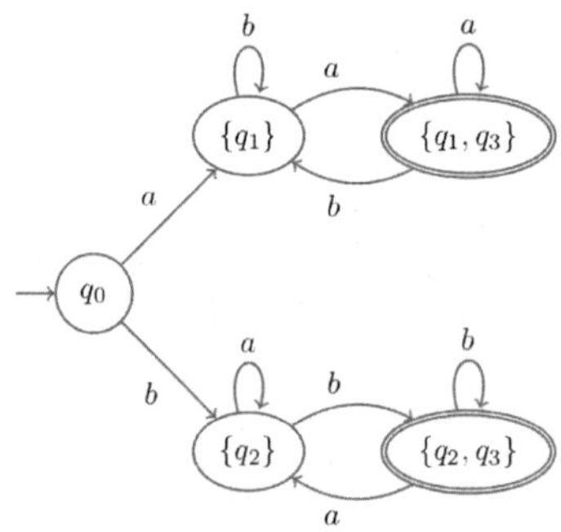

Figure 2–21: DFA for beginning and ending with the same symbol

Whenever a composite state contains any accepting state from the original NFA, that state becomes an accepting state in the resulting DFA. Thus, both states containing q_3 in the figure above are accepting states.

It is usually easier to track multiple moves in a NFA with a transition table instead of a tree. The following is a transition table for the NFA in Figure 2–15.

State	*a*	*b*
q_0	q_1	q_2
q_1	$\{q_1, q_3\}$	q_1
q_2	q_2	$\{q_2, q_3\}$
$+q_3$	ϕ	ϕ

Table 2–2: The NFA in Figure 2-15 as a Transition Table

The sets representing the possible output states from q_1 with an *a* and from q_2 with a *b*. There are no valid moves from state q_3, indicated by the empty set.

To begin the conversion process to a DFA, we populate a *new* transition table, listing only the start state at first. We refer to Table 2-2 to determine where the NFA takes us from state q_0 with each symbol of the alphabet, as shown in Table 2–3.

State	*a*	*b*
q_0	$\{q_1\}$	$\{q_2\}$

Table 2–3: Starting the NFA-to-DFA conversion process

The NFA reached states q_1 and q_2, so we add rows for these to see where they in turn take us with the possible input symbols:

State	a	b
q_0	$\{q_1\}$	$\{q_2\}$
$\{q_1\}$		
$\{q_2\}$		

Table 2–4: Adding the two newly reached states

As we track where we can move from these states, we update our working table:

State	a	b
q_0	$\{q_1\}$	$\{q_2\}$
$\{q_1\}$	$\{q_1, q_3\}$	$\{q_1\}$
$\{q_2\}$	$\{q_2\}$	$\{q_2, q_3\}$

Table 2–5: Tracking the outputs from states q_1 and q_2

We have obtained two new "states," $\{q_1, q_3\}$ and $\{q_2, q_3\}$, which we process in Table 2–6 below.

State	a	b
q_0	$\{q_1\}$	$\{q_2\}$
$\{q_1\}$	$\{q_1, q_3\}$	$\{q_1\}$
$\{q_2\}$	$\{q_2\}$	$\{q_2, q_3\}$
$+\{q_1, q_3\}$	$\{q_1, q_3\}$	$\{q_1\}$
$+\{q_2, q_3\}$	$\{q_2\}$	$\{q_2, q_3\}$

Table 2–6: The final DFA

No new states have appeared in the body of our table at this point, so we are done. Having tracked all possible movements explicitly, there is no ambiguity of moves using the new set of (possibly composite) states, so the result is *deterministic*, and is equivalent to the transition diagram in Figure 2–21. The technique we have just illustrated is known as *subset construction*[3] (because our destination "states" can represent subsets of the original set of states).

Example 2–14

Let's now convert the NFA in Figure 2–18 to a DFA, referring to its transition table below.

State	a	b	c
q_0	$\{q_0, q_1, q_2\}$	ϕ	ϕ
q_1	ϕ	ϕ	$\{q_1, q_2\}$
$+q_2$	ϕ	$\{q_2\}$	ϕ

Table 2–7: Transition table for the NFA in Figure 2–18

As explained earlier, the output transitions for each state yield a subset of $\{q_0, q_1, q_2\}$. Because of the lambda transition, there is a free ride to q_2 whenever q_1 is reached. The set of states immediately reachable upon entering a state is called its **lambda closure**, which includes the state itself along with

[3]It is also known as the Rabin-Scott *powerset construction*, named after the authors of a 1959 paper.

all states reachable from that state by one or more lambda-transitions (so the lambda closure of q_1 in Figure 2–19 is the set $\{q_1, q_2\}$). Whenever you enter a state, you automatically reach the other states in its lambda closure without consuming a symbol. For this example, q_1, therefore, **will never appear** as an output state inside the body of a working transition table **without** q_2.

Definition 2.4

The **lambda closure** of a state is the set containing the state itself along with all other states reachable from that state via one or more consecutive lambda transitions (consuming no input).

As before, we begin the conversion to a DFA by constructing a transition table with only the start state of the original NFA[4], and see where it takes us with each input symbol. (See Table 2–8.)

State	*a*	*b*	*c*
q_0	$\{q_0, q_1, q_2\}$	ϕ	ϕ

Table 2–8: First step in the NFA-to-DFA conversion

As expected, with an a the machine can reach q_0, q_1 and q_2 (the latter because of the lambda edge from q_1 to q_2) from state q_0, so we track these states together as a composite state. We will add this to our running list of states. The state ϕ in this working table represents the fact that there are no edges for the corresponding symbols and states in the original machine, and will become a jail state in the resulting DFA. We won't bother to add a row for a jail state since we know that jails just loop on every symbol.

To fill in the second row, we must track the behavior of all three states inside the composite state simultaneously. From there, an a again reaches the composite state $\{q_0, q_1, q_2\}$. With a c as input, however, state q_1 moves to itself, and, because of the free ride, to q_2 also. (See Table 2–9)

State	*a*	*b*	*c*
q_0	$\{q_0, q_1, q_2\}$	ϕ	ϕ
$\{q_0, q_1, q_2\}$	$\{q_0, q_1, q_2\}$	$\{q_2\}$	$\{q_1, q_2\}$

Table 2–9: Processing the new composite state $\{q0, q1, q2\}$

We have reached two new states, so we will add two more rows to track them. See Table 2–10.

[4]Because the start state may have outgoing lambda transitions, the start state for the DFA may be a *set* of states. That is not the case here, but the first step in this construction is always to use the lambda closure of the NFA's start state as the start state of its DFA.

State	a	b	c
q_0	$\{q_0, q_1, q_2\}$	ϕ	ϕ
$+\{q_0, q_1, q_2\}$	$\{q_0, q_1, q_2\}$	$\{q_2\}$	$\{q_1, q_2\}$
$+\{q_2\}$	ϕ	$\{q_2\}$	ϕ
$+\{q_1, q_2\}$	ϕ	$\{q_2\}$	$\{q_1, q_2\}$

Table 2–10: Complete Transition Table for the DFA.

No new states have appeared, so we finish by indicating the accepting states. The transition diagram for the complete DFA appears in Figure 2–22. Note the jail state.

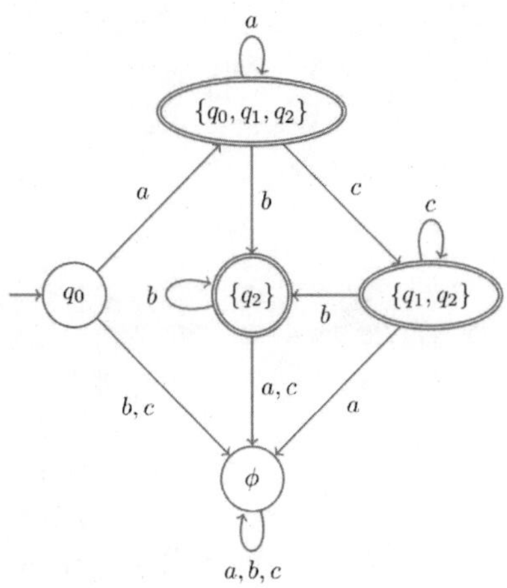

Figure 2–22: The complete DFA diagram after the conversion.

Initial states can have lambda transitions, so remember to make the initial state of the new DFA the lambda closure of the initial state of the NFA when converting an NFA to a DFA.

The following steps summarize the construction of a DFA from a NFA:

1. Initiate a working transition table for the resulting DFA with the lambda closure of the start state of the NFA as the start state of the DFA.
2. For each unprocessed (possibly composite) state, S, in the list of states in the left column:
   - For each state, s, in S:
   - For each symbol, c, in the alphabet:
   - For each state, q, reachable from s by c:
   - Add the states of the lambda closure of q to the cell in row S and column c.

3. For each new (possibly composite) state obtained, add it to the list of states to process in the left column; go to step 2 unless no unprocessed (possibly composite) state remains in the left column.
4. All resulting states containing any accepting state from the original NFA are accepting states in the DFA.

Since any NFA has an equivalent DFA accepting the same language, both types of automata characterize the class of regular languages.

NFAs and Complements

Recall that the complement of a regular language can be recognized by a machine obtained by inverting the acceptability of each state in a DFA that recognizes the language. The same strategy does not apply, however, to finding the complement of a regular language using a NFA. Why not?

One reason is because NFAs often omit possible moves for input strings, since they represent only what is acceptable in the language. When complementing a language, we are interested in what is not acceptable in the original language, so starting with a NFA won't work.

Example 2–15

Consider the language $\{a, bab\}^*$, which the following NFA accepts.

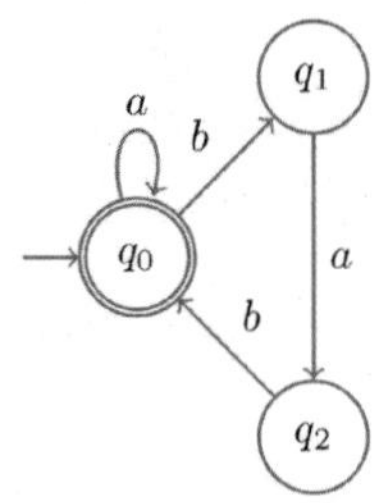

Figure 2–23: NFA for a, bab^*

Were we to attempt to complement this machine by only reversing the acceptability of the states above, we would get the NFA in Figure 2-24:

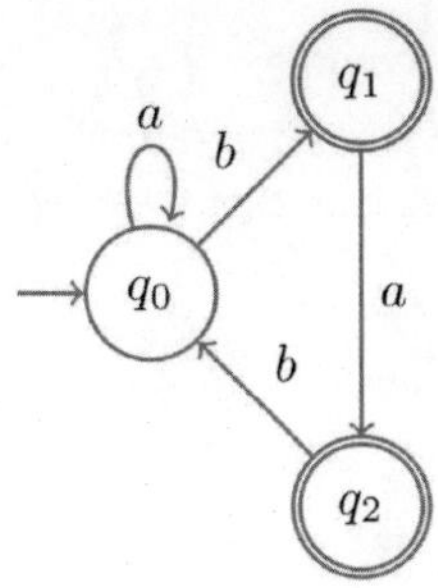

Figure 2–24: Inverting the states in Figure 2–23

This NFA does not accept the string *bb*, which is certainly in the complement of $\{a, bab\}^*$. To recognize the complement, we must start with a **DFA** that accepts the language, which in this case requires only adding a jail state for the missing moves. (See Figure 2–25.)

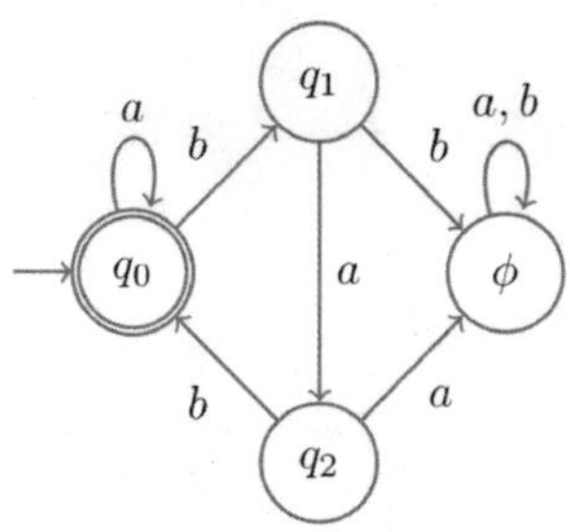

Figure 2–25: DFA for a, bab^*

We can now invert each state's acceptability to get a DFA for the complement of the language. You can see that the DFA in Figure 2–26 accepts the string *bb*.

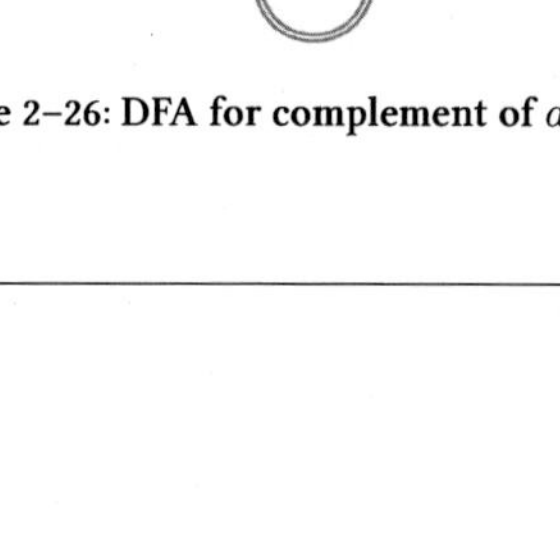

Figure 2–26: DFA for complement of a, bab^*

Example 2–16

Even if a NFA does not omit any possible moves, inverting its state acceptability can still fail to give the correct complement. Consider the NFA in Figure 2–27, which recognizes the language aa^*b^*.

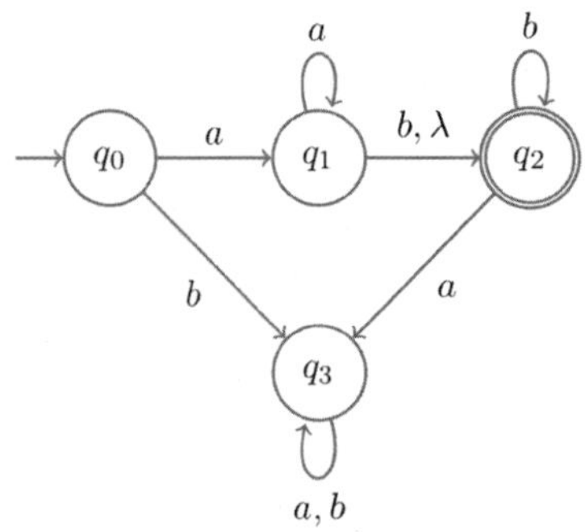

Figure 2–27: NFA accepting aa^*b^*

This NFA is not missing any out-edges, but it has a lambda edge from state q_1 to state q_2. It accepts the singleton string a, as it should. Now look at the NFA that results from inverting the acceptability of each state in Figure 2–28.

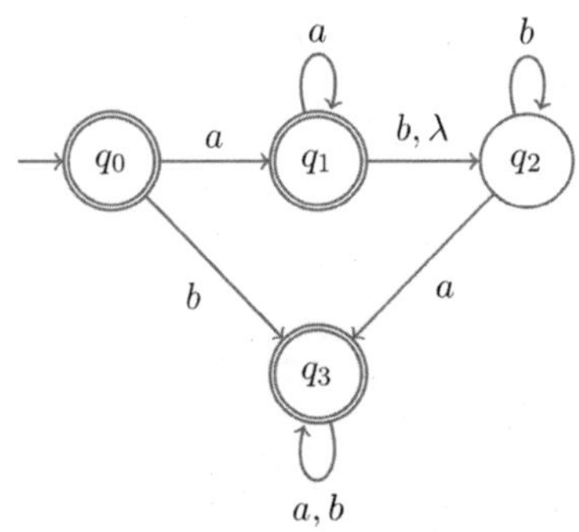

Figure 2–28: Inverting Figure 2–27

This NFA also accepts the string a, so it is not the complement after all.

To find the complement of a regular language, start with a **DFA**.

Key Terms

non-determinism • lambda transition • lambda closure • composite state • subset construction

Exercises

1. List all possible paths in the automaton in Figure 2–16 (Example 2–11) that accept the string $baaababba^*$.
2. Draw a NFA for the language of all strings over $\Sigma = \{a, b\}$ that begin and end with the same symbol, (similar to example 2–10), but without the restriction that the string is at least length 2—that is, the strings λ, a, b should also be accepted.

The following NFA pertains to problems 3–5.

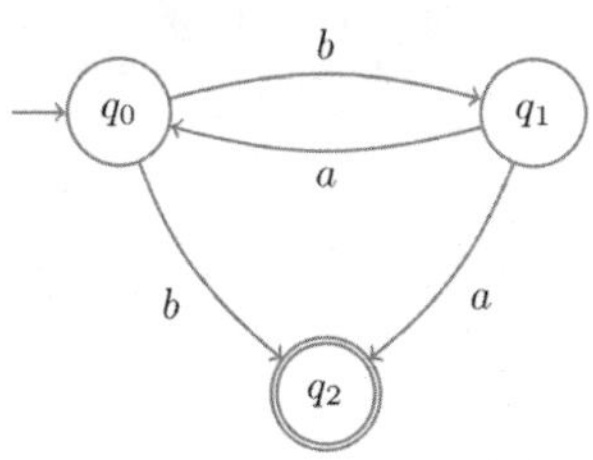

3. What language does this NFA represent?
4. Convert this NFA to a DFA using the tabular method shown in this section.
5. Draw the transition graph for the DFA from the previous problem.
6. Draw a NFA that accepts all strings over $\Sigma = \{a, b\}$ that either end in *ab* or contain the substring *bb*.
7. Convert the NFA in the previous exercise to a DFA. Draw the transition diagram.
8. Draw a NFA that accepts all strings over $\Sigma = \{a, b\}$ that have both *aa* and *bb* as substrings.
9. Convert the following NFA to a DFA. Draw the transition diagram.

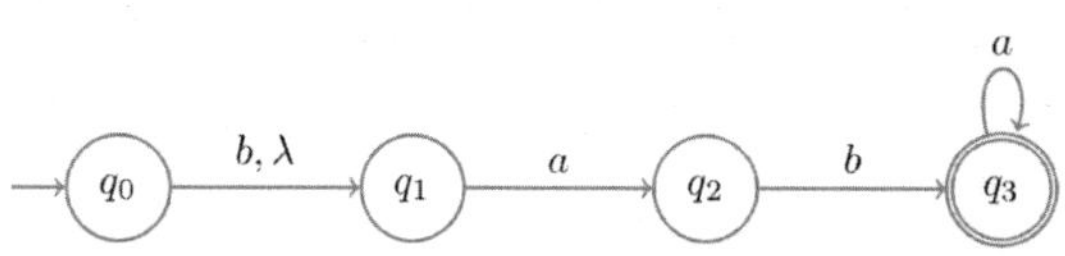

10. Draw a NFA that accepts the language ab^*, (i.e., the language $\{a, ab, abb, abbb, ...\}$). Then find its complement and draw its transition graph.

2.3 Minimal Automata

DFAs are useful in recognizing text patterns and they are easily implemented in code. We haven't considered, however, whether an automaton is "efficient." Our measure of efficiency is the *number of states*, since that determines the number of possible transitions that exist.

Example 2–17

Examine the following DFA for any redundant states.

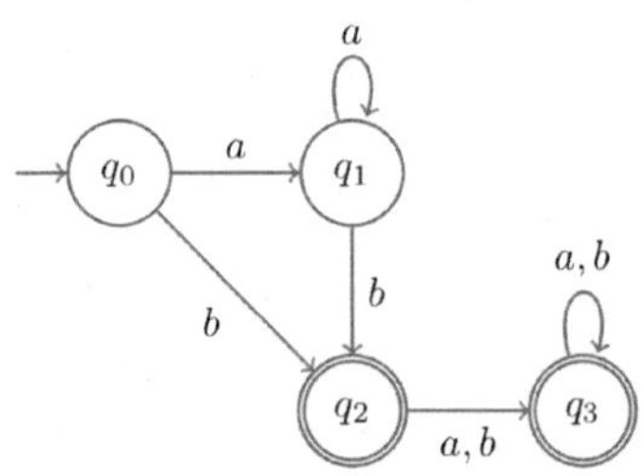

Figure 2–29: Is this DFA "efficient"?

Once we reach state q_2, the machine remains in an accepting state regardless of subsequent input. States q_2 and q_3 can therefore collapse into a single, accepting state, as shown in the next figure.

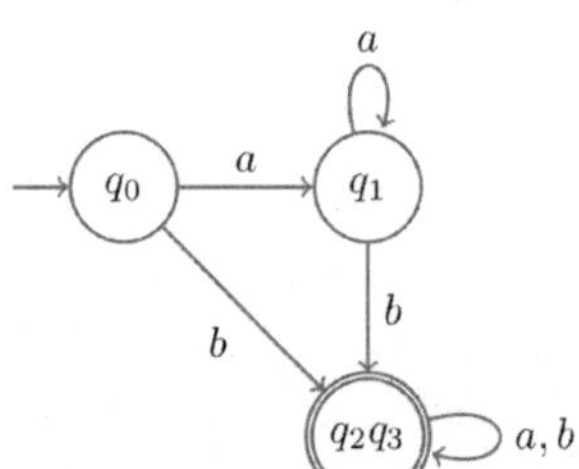

Figure 2–30: Combining states q_2 and q_3

Now think about what language this machine accepts. If a string starts with a *b*, it is accepted. If it starts with an *a*, it is not accepted until a *b* is read. In other words, this machine accepts the language of strings that have at least one *b*. We can therefore combine states q_0 and q_1 to obtain the minimal DFA in Figure 2–31.

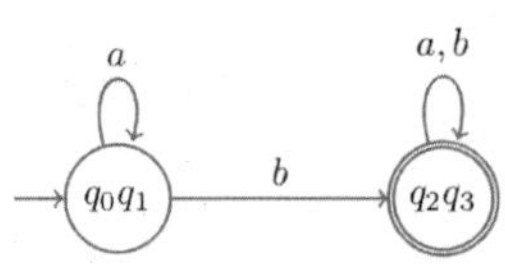

Figure 2–31: A minimal DFA for the language accepted in Figure 2–29

We have discovered that states q_0 and q_1 play the same role in the original DFA. States q_2 and q_3 are likewise indistinguishable in their function, and we have reduced the number of states from 4 to 2.

Example 2–18

Which states are indistinguishable in the following DFA?

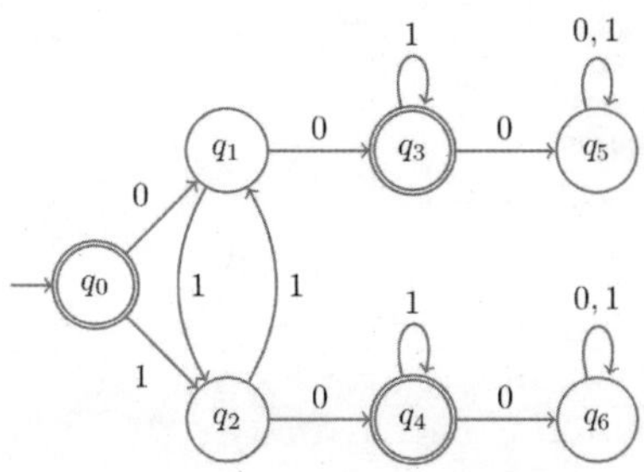

Figure 2–32: Another inefficient DFA

First, note that there are two jails: states q_5 and q_6. There is never a need for more than one jail state, so we combine those. A little thought reveals that q_3 and q_4 behave identically, and q_1 and q_2 as well. The minimal DFA appears in Figure 2–33 below.

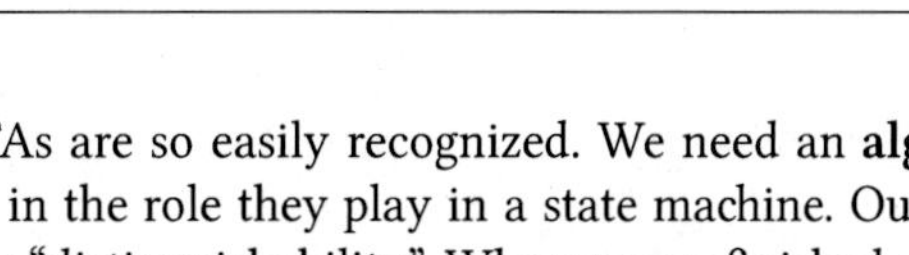

Figure 2–33: Minimal DFA for Figure 2–32

Not all redundancies in DFAs are so easily recognized. We need an **algorithm** to determine when states are indistinguishable in the role they play in a state machine. Our approach is to examine all possible pairings of states for "distinguishability". When we are finished, we can combine into a single state those states that have not been distinguished from each other.

We know that final states clearly play a different role than non-final states, since they give different "output" (accept vs. reject), so we first mark all final/non-final pairs of states as *distinguishable*. It is convenient to use a tabular approach to track the process. We will use the DFA in Figure 2–29 for illustration. Examine the structure of the following table.

	q_1	q_2	q_3
q_0		✔	✔
q_1		✔	✔
q_2			

Table 2–11: Tracking distinguishable states in Figure 2–29

The six unshaded cells in the upper triangle in the body of the table correspond to the six possible pairings of states in the DFA: (q_0, q_1), (q_0, q_2), (q_0, q_3), (q_1, q_2), (q_1, q_3) and (q_2, q_3). We have placed check marks in the cells that pair a final state with a non-final state, since we know that they are distinguishable from the get-go. This is the initial configuration for the minimization process.

We now visit the two remaining unchecked cells to see if we can distinguish them or not. We define distinguishability of states recursively as follows.

Definition 2.5

Two DFA states, q_1 and q_2, are **distinguishable** if:

- **exactly one** of the states is an *accepting state*, or
- The states $\delta(q_1, c)$ and $\delta(q_2, c)$ are *distinguishable* for *at least one symbol* $c \in \Sigma$

In other words, two states are distinguishable if they have different acceptability or if they move with the same symbol to states which have been previously found to be distinguishable (so the minimization algorithm will be iterative). Distinguishable states transition to semantically different subsequent states with the *same input symbol.*

We can now check (q_0, q_1) and (q_2, q_3) for distinguishability. (We already know from Example 2–17 that neither pair is distinguishable, but let's illustrate the algorithm, which we will formalize shortly.) For (q_0, q_1), we have, referring to Figure 2-29:

$$\delta(q_0, a) = q_1, \delta(q_1, a) = q_1$$
$$\delta(q_0, b) = q_2, \delta(q_1, b) = q_2$$

The output states for each input symbol state are identical, so they are certainly indistinguishable. Looking at (q_2, q_3) we find the following:

$$\delta(q_2, a) = q_3, \delta(q_3, a) = q_3$$
$$\delta(q_3, b) = q_3, \delta(q_3, b) = q_3$$

Both inputs move to q_3, so these states can also be combined, as we surmised when we arrived at the minimal DFA in Figure 2–31.

Example 2–19

Now let's use the process to minimize the DFA in Figure 2–32. We first initialize the table, marking pairs of final vs. non-final states distinguishable.

	q_1	q_2	q_3	q_4	q_5	q_6
q_0	✔	✔			✔	✔
q_1			✔	✔		
q_2			✔	✔		
q_3					✔	✔
q_4					✔	✔
q_5						

Table 2–12: Initial table for minimizing Figure 2–32

We have nine pairs corresponding to the empty cells above to test for distinguishability:

(q_0, q_3):

$\delta(q_0, 0) = q_1,\ \delta(q_3, 0) = q_5$

$\delta(q_0, 1) = q_2,\ \delta(q_3, 1) = q_3$ ✔

(q_0, q_4):

$\delta(q_0, 0) = q_1,\ \delta(q_4, 0) = q_6$

$\delta(q_0, 1) = q_2,\ \delta(q_4, 1) = q_4$ ✔

(q_1, q_2):

$\delta(q_1, 0) = q_3,\ \delta(q_2, 0) = q_4$

$\delta(q_1, 1) = q_2,\ \delta(q_2, 1) = q_1$

(q_1, q_5):

$\delta(q_1, 0) = q_3,\ \delta(q_5, 0) = q_5$ ✔

(q_1, q_6):

$\delta(q_1, 0) = q_3,\ \delta(q_6, 0) = q_6$ ✔

(q_2, q_5):

$\delta(q_2, 0) = q_4,\ \delta(q_5, 0) = q_5$ ✔

(q_2, q_6):

$\delta(q_2, 0) = q_4,\ \delta(q_6, 0) = q_6$ ✔

(q_3, q_4):

$\delta(q_3, 0) = q_5,\ \delta(q_4, 0) = q_6$

$\delta(q_3, 1) = q_3,\ \delta(q_4, 1) = q_4$

(q_5, q_6):

$\delta(q_5, 0) = q_5,\ \delta(q_6, 0) = q_6$

$\delta(q_5, 1) = q_5,\ \delta(q_6, 1) = q_6$

When we find moves that distinguish two states, we don't have to check any remaining symbols for that pair (as with the last four check marks above). During this round, we found six new pairs of distinguishable states, so we update our table as follows.

	q_1	q_2	q_3	q_4	q_5	q_6
q_0	✔	✔	✔	✔	✔	✔
q_1			✔	✔	✔	✔
q_2			✔	✔	✔	✔
q_3					✔	✔
q_4					✔	✔
q_5						

Table 2-13: Round 2 of checking for distinguishable pairs

Since we found new check marks, we must process the remaining empty cells again, in case the newfound distinguishable pairs lead to yet other ones. Doing so, however, we find that no new changes occur, so we now combine (q_1, q_2), (q_3, q_4), and (q_5, q_6) to obtain Figure 2–33.

We can now describe the minimization algorithm as follows:

1. For a machine with n states, form a $(n-1) \times (n-1)$ triangular table, listing the first $n-1$ states down the left side as row headers, and the last $n - 1$ states horizontally as column headers.
2. Initialize the table by placing a check mark in those cells where **exactly one** of the corresponding states in the pair is a final state.
3. For each remaining empty cell: if the two corresponding states move to distinguishable states on the **same symbol** of the alphabet, place a check mark in that cell.
4. If a new check mark was entered anywhere in step 3, repeat step 3.
5. Combine the states that correspond to any remaining empty cells.

Indistinguishability (but not distinguishability!) is transitive: if the pair (x, y) is indistinguishable, as well as (y, z), then the pair (x, z) is also an indistinguishable pair, and all three states can be combined into one state, $\{x, y, z\}$.

Example 2–20

It was easy to see which states could be combined in the previous two examples. The DFA in in the next figure isn't all that easy to minimize by cursory inspection.

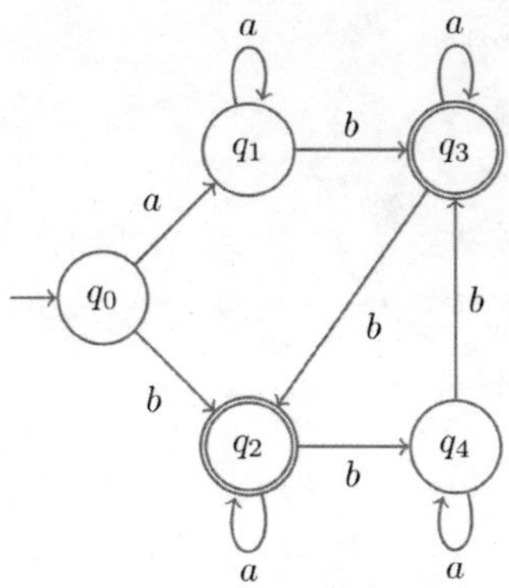

Figure 2–34: Another DFA to minimize

The initial setup for the minimization process follows in Table 2-14.

	q_1	q_2	q_3	q_4
q_0		✔	✔	
q_1	■	✔	✔	
q_2	■	■		✔
q_3	■	■	■	✔

Table 2-14: Initial table for minimizing the DFA in Figure 2–34

We now process each empty cell:

(q_0, q_1):

$$\delta(q_0, a) = q_1,\ \delta(q_1, a) = q_1$$
$$\delta(q_0, b) = q_2,\ \delta(q_1, b) = q_3$$

(q_0, q_4):

$$\delta(q_0, a) = q_1,\ \delta(q_4, a) = q_4$$
$$\delta(q_0, b) = q_2,\ \delta(q_4, b) = q_3$$

(q_1, q_4):

$$\delta(q_1, a) = q_1,\ \delta(q_4, a) = q_4$$
$$\delta(q_1, b) = q_3,\ \delta(q_4, b) = q_3$$

(q_2, q_3):

$$\delta(q_2, a) = q_2,\ \delta(q_3, a) = q_3$$
$$\delta(q_2, b) = q_4,\ \delta(q_3, b) = q_2 \quad ✔$$

On this iteration, we see that states q_2 and q_3 are distinguishable, so we update the table in preparation for another iteration.

	q_1	q_2	q_3	q_4
q_0		✔	✔	
q_1		✔	✔	
q_2			✔	✔
q_3				✔

Table 2-15: Distinguishing states q_2 and q_3

Now we test for other distinguishable pairs based on this new information:

(q_0, q_1):

$$\delta(q_0, a) = q_1,\ \delta(q_1, a) = q_1$$
$$\delta(q_0, b) = q_2,\ \delta(q_1, b) = q_3 \quad ✔$$

(q_0, q_4):

$$\delta(q_0, a) = q_1,\ \delta(q_4, a) = q_4$$
$$\delta(q_0, b) = q_2,\ \delta(q_4, b) = q_3 \quad ✔$$

(q_1, q_4):

$$\delta(q_1, a) = q_1,\ \delta(q_4, a) = q_4$$
$$\delta(q_1, b) = q_3,\ \delta(q_4, b) = q_3$$

This leaves only one indistinguishable pair, (q_1, q_4). A quick check verifies that these states remain indistinguishable, so we combine them in the final result in Figure 2-35.

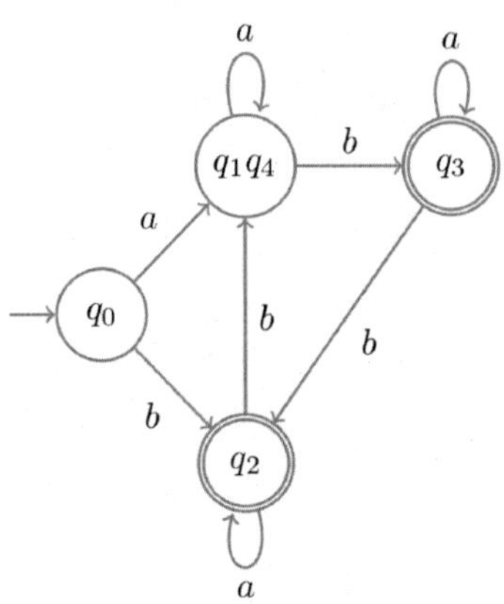

Figure 2–35: Minimal DFA for Figure 2–34

Key Terms

minimal DFA • distinguishable states • transitive

Exercises

1. Show that the DFA in Figure 2–9 is minimal.
2. Minimize the DFA defined by the transition function below. The accepting states are $\{q_3, q_4, q_8, q_9\}$.

$$\delta(q_0, a) = q_1, \delta(q_0, b) = q_9$$
$$\delta(q_1, a) = q_8, \delta(q_1, b) = q_2$$
$$\delta(q_2, a) = q_3, \delta(q_2, b) = q_2$$
$$\delta(q_3, a) = q_2, \delta(q_3, b) = q_4$$
$$\delta(q_4, a) = q_5, \delta(q_4, b) = q_8$$
$$\delta(q_5, a) = q_4, \delta(q_5, b) = q_5$$
$$\delta(q_6, a) = q_7, \delta(q_6, b) = q_5$$
$$\delta(q_7, a) = q_6, \delta(q_7, b) = q_5$$
$$\delta(q_8, a) = q_1, \delta(q_8, b) = q_3$$
$$\delta(q_9, a) = q_7, \delta(q_9, b) = q_8$$

3. Find a minimal DFA for the following NFA.

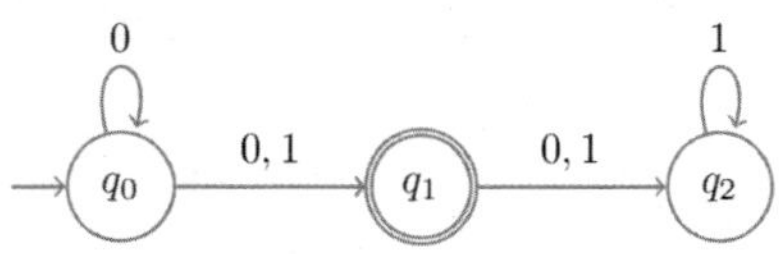

2.4 Machines with Output

Example 1–4 showed a finite automaton that echoed all its input except for the text of dollar-delimited comments. The generic term for such an output-producing machine is *finite-state transducer*, borrowing the name from electronics (a transducer converts energy from one form to another). Our transducers convert input text to output text. The machine in Figure 2–36 below emits output as it moves from one state to another, using a slash on transitions to separate the input from the output. Such a transducer is also called a Mealy machine[5].

Example 2-1

The following Mealy machine emits a 1 whenever the input symbol changes, and a 0 otherwise.

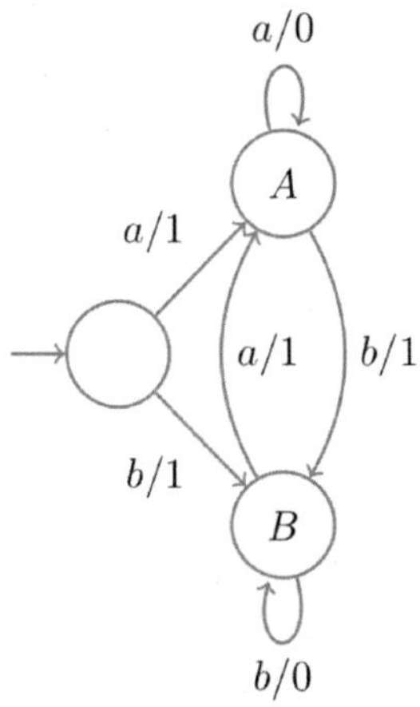

Figure 2–36: A Mealy machine that tracks changes in input

The first symbol always emits a 1, since it follows nothing (which is considered a "change"). Given the input string *aababba* the output is 1011101. The states are aptly named to reflect the most recent symbol read.

Example 2-22

Mealy machines can act as *recognizers* for specific substrings by printing a 1 whenever they encounter them. This has an advantage over a DFA recognizer because it indicates the position of the last symbol of multiple, matching substrings. The machine in Figure 2-37 recognizes the substrings aa and bb.

[5]Named after George H. Mealy who introduced them to describe sequential circuits. Another type of machine, named after Edward F. Moore, emits output when a state is entered instead of during a transition. Both types of machines solve the same set of problems; one may be preferable over the other in different situations. We will use only Mealy machines.

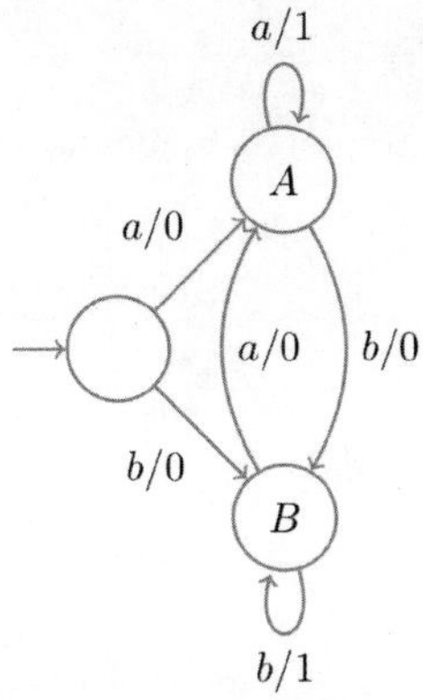

Figure 2-37: Recognizes substrings *aa* or *bb*

The output for the input string $aaabbabbba$ is 0110100110. This machine catches overlapping substrings, hence the initial substring aaa above prints 011.

Example 2–23

A parity bit is a bit that is added at the end of a bit string to indicate whether the number of 1-bits in the original string is even or odd. An *even parity bit* of 0 will be appended if the number of original 1-bits is even; otherwise the even parity bit is 1. Using an even parity bit is a rudimentary error-checking process that always produces an even number of 1-bits in a transmitted string. The machine in the following figure appends an even parity bit to its input.

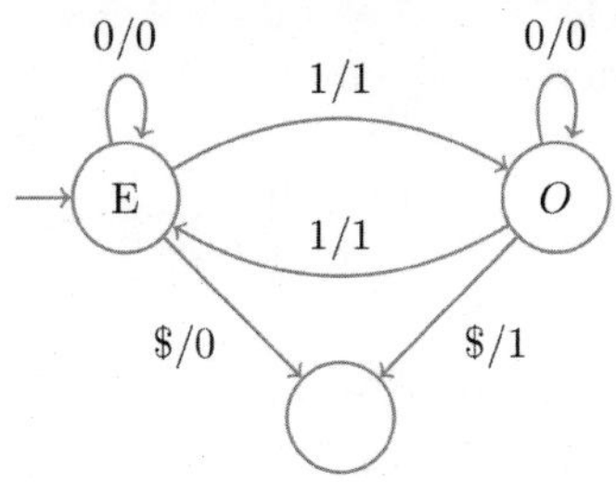

Figure 2–38: An even-parity checker

It is necessary to know when the string ends, so we use a dollar sign to mark the end of the input string.

Example 2–24

The following diagram from Cohen[6], represents a sequential circuit.

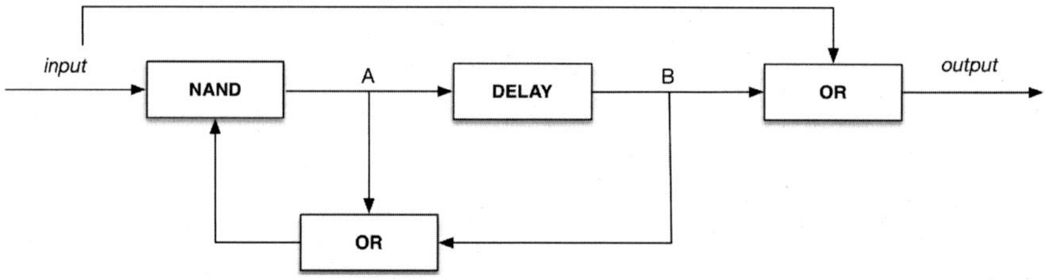

Figure 2–39: A simple sequential circuit

The input bit feeds into both the first NAND and the final OR. At each clock pulse the circuit advances, so the new B becomes what the old A was, the new A becomes the NAND of the input with the OR of the old A and old B, and the OR of the old B and the input is the output for that pulse. The internal state is the pair of values in A and B at the same moment in time, namely, one of $\{(0,0),(0,1),(1,0),(1,1)\}$. We can trace the possible configurations in a transition table. We begin with the start state, and see where it leads (because not all possibilities of combinations will necessarily appear in a given circuit). (See Table 2–17).

State	0	1
q_0: **(0, 0)**		

Table 2–17: Start state of the sequential circuit

Now examine what happens in state q_0 (i.e., where $A = B = 0$) and the input is 0. The new B is the old A, which is 0. The new A is $\neg(0 \wedge (0 \vee 0)) = \neg 0 = 1$. The output is $0 \vee 0 = 0$. If the input is 1, then A becomes $\neg(1 \wedge (0 \vee 0))$, which is 1, B is still 0, but the output is 1 this time. The table below updates its first row with these results, using $q_1 = (1,0)$.

State	0	1
q_0: **(0, 0)**	(1,0)/0	(1,0)/1
q_1: **(1, 0)**		

Table 2–18: First row of Table 2–17 completed

We now seek where $q_1(A = 1, B = 0)$ leads. If the input is 0, the new A becomes $\neg(0 \wedge (1 \vee 0)) = \neg 0 = 1$, as before, and, as always, B becomes the old A, so we go to a new state, $(1,1)$. If the input is 1, the new A is $\neg(1 \wedge (1 \vee 0)) = 0$, and $B = 1$, leading to a new state, $(0,1)$:

[6]Cohen, Daniel, *Introduction to Computer Theory*, Second Edition, Wiley, 1997, p. 162.

State	0	1
q_0: $(0,0)$	(1,0)/0	(1,0)/1
q_1: $(1,0)$	(1,1)/0	(0,1)/1
q_2: $(0,1)$		
q_3: $(1,1)$		

Table 2–19: Second row completed

If you complete the table, you should get a Mealy machine equivalent to the following diagram.

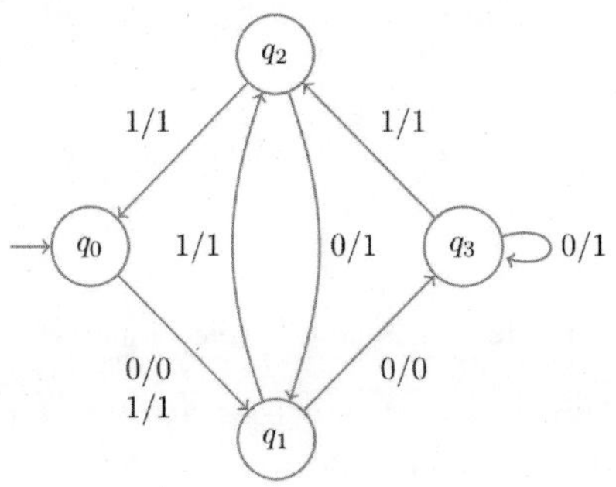

Figure 2–40: Mealy machine for the circuit in Figure 2-40

Computer Arithmetic

Mealy machines naturally model low-level computer arithmetic algorithms. A one's-complement machine, for example, is modeled by the following, trivial machine.

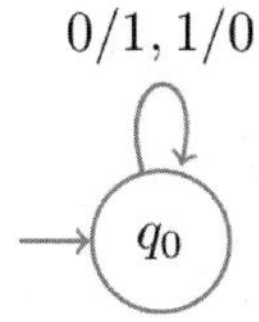

Figure 2–41: A one's-complement machine

Example 2–25

Adding two binary numbers is also "Mealy-able," but requires reading corresponding bits simultaneously, right-to-left. The machine will therefore take *pairs* of bits as input, in reverse order. Recall how to add binary numbers by hand:

1101
<u>0110</u>
10011

The Mealy machine below considers each vertical bit-pair, right-to-left, mimicking what we do manually. We assume it emits the result in right-to-left order.

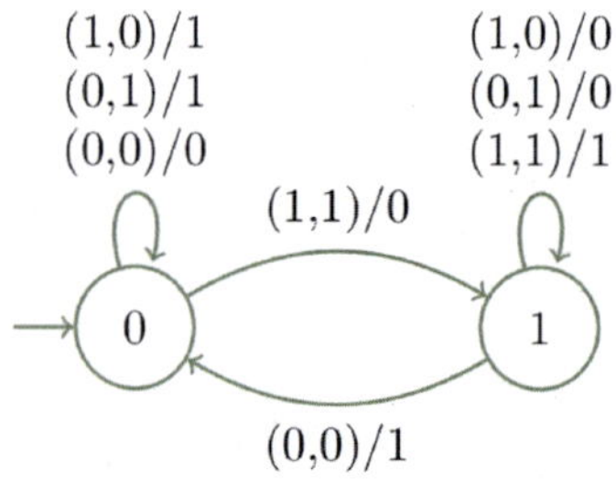

Figure 2–42: Adding two binary numbers

The one thing missing is the leftmost carry bit (only 0011 is printed for the sum above instead of 10011). To solve this problem, we assume that each state emits the bit in its label, which is the carry, when the machine halts there. The code below implements the carry.

```
def binadd(x,y):
    assert len(x) == len(y)
    state = "0"
    pairs = zip(x,y)            # combine strings pairwise
    result = ""
    for b1,b2 in reversed(pairs):
        if state == "0":
            if b1=="1" and b2=="0" or b1=="0" and b2=="1":
                result += "1"
            elif b1=="0" and b2=="0":
                result += "0"
            elif b1=="1" and b2== "1":
                result += "0"
                state = "1"
            else:
                assert False, "invalid input"
        else:   # state == "1"
            if b1=="1" and b2=="0" or b1=="0" and b2=="1":
                result += "0"
            elif b1=="1" and b2=="1":
                result += "1"
            elif b1=="0" and b2== "0":
                result += "1"
```

```
            state = "0"
        else:
            assert False, "invalid input"
    if state == "1":
        result += "1"               # append carry
    return result[::-1]             # reverse result

print(binadd("1101","0110"))        # 10011
```

Example 2–26

Subtracting two binary numbers follows a similar pattern, except the states track what is *borrowed* instead of what is carried:

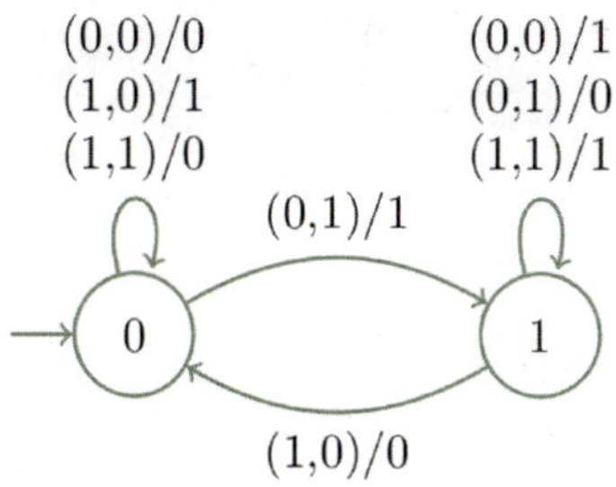

Figure 2–43: Subtracting binary numbers

Using the same input as in the previous example, we get 1101-0110=0111, which in decimal is 13-6=7. Subtracting the other way, we get 0110-1101=1001. The latter result is the two's-complement of 0111, and represents -7 decimal, as expected. Therefore, the bit in the label in each state is the *borrow* (or equivalently, the *sign-bit*; 1 for negative, 0 otherwise).

Example 2–27

As a final example of computer arithmetic, we design a Mealy machine that rounds its binary input down to nearest even number. All this requires is to make the last bit zero if it isn't already. But we need to know when the last symbol has been read, so we need an *end-of-string symbol*; we'll use the dollar sign again. Furthermore, we need to have a *one-symbol delay* in emitting output; since we don't print the dollar sign, we always print the previous symbol consumed after reading the current character. See Figure 2–44 below.

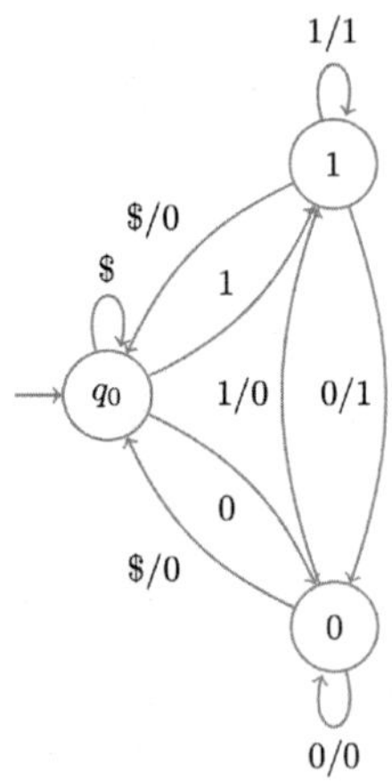

Figure 2–44: Rounds to the nearest even number

The outgoing edges from q_0 emit nothing. Instead, we use the other two states to track the previous symbol read. When the string has been exhausted, we emit a 0 no matter what. The output for 101$ is 100, as expected.

Lexical Analysis

Mealy-style machines are useful in separating input into meaningful parts. In the next example, we consider how to design a machine that will ignore C-style comments from its input, while outputting everything else.

Example 2–28

C-style comments are supported by many popular programming languages and are delimited with a "/*" at the beginning and a "*/" at the end. Since the delimiter is a two-symbol sequence, we need to detect when a slash and asterisk occur *together*. Reading a slash may or may not introduce a comment. What if a slash is followed by another slash? The second slash may or may not belong to a comment, but the first slash must pass through as output regardless. The following diagram is a start on the design for a Mealy Machine to solve the problem.

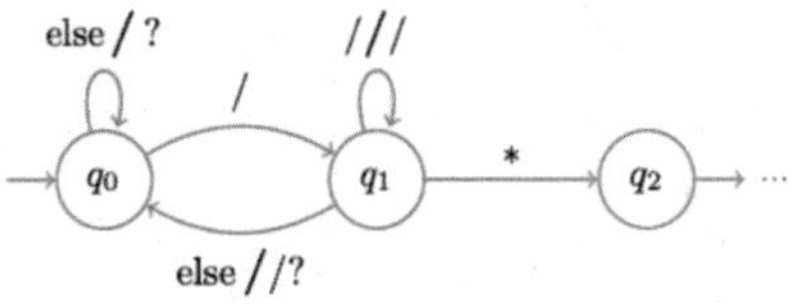

Figure 2–45: Partial machine for removing C-style comments

Technically, Mealy machines have single symbols for input and always have an output symbol on each edge, so that the input and output have the same number of symbols (this is important in electronics). We are obviously taking some liberties with notation here. First, we use the input "else" to cover all cases where the other edges leaving a state do not apply. We also use the question mark as a wildcard symbol representing "the current symbol being processed," as we did in Chapter 1. Finally, we omit the separating slash when a transition produces no output, as in the edges from q_0 to q_1 and q_1 to q_2 above. (Some authors use lambda after a slash to indicate no output.)

Starting in state q_0, we merely echo the input until we encounter a slash, in which case we move to state q_1. At this point, we don't know if that slash actually begins a comment. If an asterisk does not follow immediately, we must echo the *previous* slash, so we are *one symbol behind* in emitting output in state q_2. We can also encounter multiple slashes in a row (e.g., if the code contains `x //* this is the divisor: */y`.) Even though it may not be syntactically correct for some languages, it is also possible that a language uses `//` as a valid operator (like Python 3). If a language uses C-style comments *and* has such an operator, as in `x//y`, then when we reach the `y`, we must output both the slash and the `y`. This explains why the edge going back to state q_0 from q_1 prints two symbols.

Once we get an asterisk in state q_1, we are inside a comment so we do not print the preceding slash and we move to state q_2. Similar logic applies when exiting a comment, which you can complete as an exercise.

Example 2–29

One of the first steps in compiling source code, in addition to removing comments, is to recognize the "words" in the code, that is, "tokens" such as keywords, variables, constants, and operators. A token consists not only of its text but also its type, or *class*. For example, in the expression `x+1`, the compiler needs to know that `x` is a variable, `+` is an operator, and `1` is a literal/constant. A machine can extract a stream of tokens and pass the result on to the next phase of compilation.

To illustrate, we will use a simple language from Louden[7], TINY, which contains only the following tokens:

Tokens in the TINY Language

Description	Class	Representation
Identifiers	ID	Consist of only alphabetic symbols
Keywords	KWD	Special identifiers: **if then else repeat until read write end**
Numbers	NUM	Base-10 integers, leading zeroes okay
Operators	OPR	+ – * / < > ( ) = (equality)
Semi-colon	SEMI	;
Assignment	ASGN	:=

[7]Louden, Kenneth C., Compiler Construction: Principles and Practice, PWS Publishing, 1997. Used with permission.

Comments are delimited by single, opening and closing curly braces. Here is sample TINY program:

A sample TINY program

```
{ A TINY Program
    CS 3240
}
    read x;
    if 0 < x then
        fact := 1;
        repeat
            fact := {embedded comment}fact * x;
            x := x - 1
        until x=0;
        write fact
    end
    {}
```

A lexical analysis of this program could be depicted as follows:

Classifying the tokens in the TINY program

```
KWD: read
ID: x
SEMI: ;
KWD: if
NUM: 0
OPR: <
ID: x
KWD: then
ID: fact
ASSIGN: :=
NUM: 1
SEMI: ;
KWD: repeat
ID: fact
ASSIGN: :=
ID: fact
OPR: *
ID: x
SEMI: ;
ID: x
ASSIGN: :=
ID: x
OPR: -
```

```
NUM: 1
KWD: until
ID: x
OPR: =
NUM: 0
SEMI: ;
KWD: write
ID: fact
KWD: end
```

The machine in Figure 2–46 below, doesn't precisely follow the definition of a Mealy machine, but it does indicate how to produce the required output above. It collects the symbols of a token and then outputs its text and token class when it returns back to the start state.

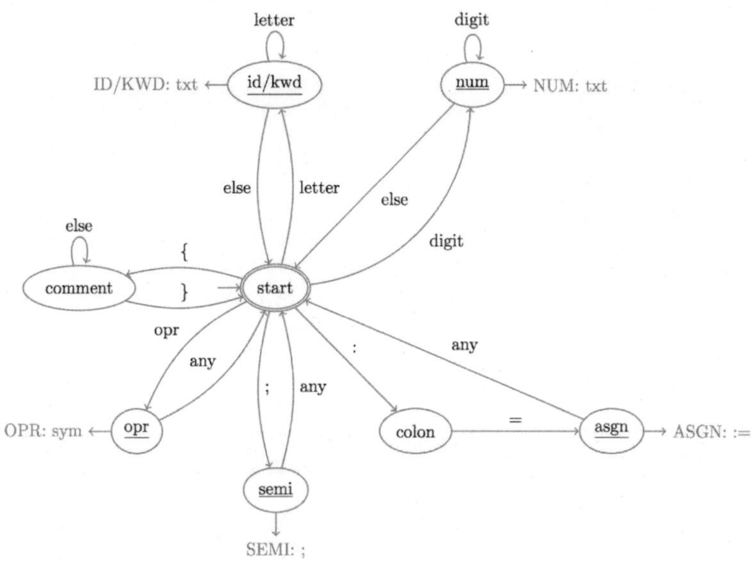

Figure 2–46: Machine that accepts TINY tokens

Note the free arrows exiting the states representing the token classes. We use these to indicate that when leaving that state, the input it has collected for its token will be emitted, along with the token class identifier. The tokens **id** and **kwd** are combined. This suggests an implementation that searches a table of keywords on the identifier to determine whether its class is ID or KWD. Because of space limitations, this diagram is missing a loop on the start state ignoring whitespace. Finally, since tokens are often delimited by non-space symbols, those symbols must be pushed back onto the input stream in some cases when returning to the start state. The start state is accepting to indicate that the input was syntactically valid. A string that does not end in the start state contains a syntax error.

Minimal Mealy Machines

We can minimize the number of states in a Mealy machine just as we did for DFAs. The machine below, which prints a 1 for each b and a 1 for every other a, starting with the first, has one more state than it needs.

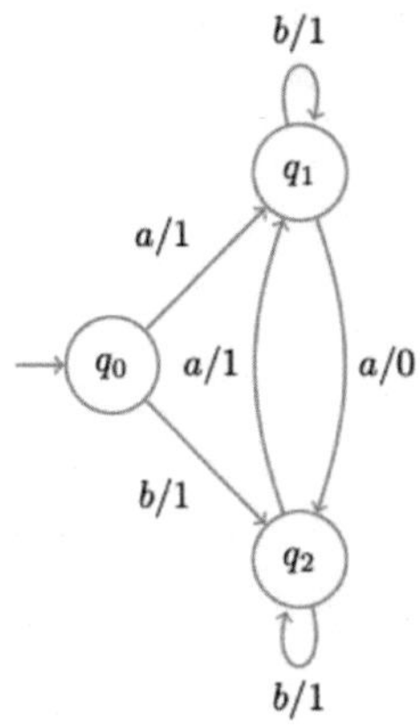

Figure 2–47: An "inefficient" Mealy Machine

Since we want the first a to emit a 1, and then alternate from there, we can eliminate state q_0 and make q_2 the initial state:

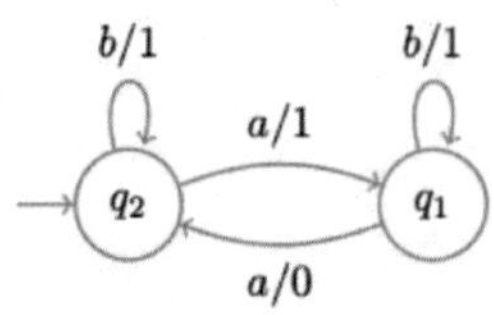

Figure 2–48: A minimal version of Figure 2–47

No move is needed for b since it emits a 1 regardless of state.

The process to minimize a Mealy machine is like the DFA minimization process in section 2.3, except instead of acceptability[8], it is the *output* that initially distinguishes one state from another. Observe the output of the different states below.

$$\delta(q_0, a) = 1, \delta(q_0, b) = 1$$
$$\delta(q_1, a) = 0, \delta(q_1, b) = 1$$
$$\delta(q_2, a) = 1, \delta(q_2, b) = 1$$

[8]Indeed, acceptability is the "output" of a state in a DFA, should the input terminate there. So the first step for both DFAs and Mealy machines is in essence to distinguish states by their "output".

The outputs for q_0 and q_2 are identical. In this simple case, we already know we can combine these states, but in general we must wait until all distinguishable pairs of states are found before merging indistinguishable states. The rest of the algorithm is identical to the state-minimization algorithm in the previous section.

To *initialize* the process of minimizing a Mealy machine, compare the *output* from different states.

Example 2–30

To illustrate the algorithm of minimizing a Mealy machine, inspect Figure 2–49 below.

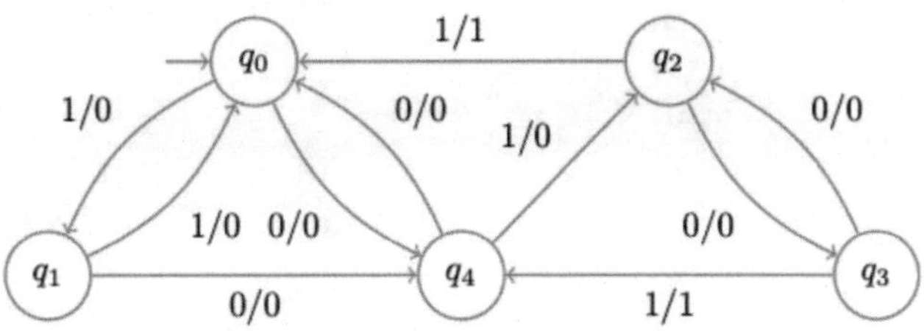

Figure 2–49: A Mealy machine to minimize

As before, we determine which states can be distinguished by first examining their output symbols:

$$\delta(q_0, a) = 0, \delta(q_0, b) = 0$$
$$\delta(q_1, a) = 0, \delta(q_1, b) = 1$$
$$\delta(q_2, a) = 0, \delta(q_2, b) = 1$$
$$\delta(q_3, a) = 0, \delta(q_3, b) = 0$$
$$\delta(q_4, a) = 0, \delta(q_4, b) = 0$$

States q_0, q_1, and q_4 each give an output of 0 for both inputs, and the other two states yield 0 for a and 1 for b. Thus, each state in $\{q_0, q_1, q_4\}$ is distinguishable from those in $\{q_2, q_3\}$, giving the initial table below.

	q_1	q_2	q_3	q_4
q_0		✔	✔	
q_1		✔	✔	
q_2				✔
q_3				✔

We now follow the state minimization algorithm from section 2.3:

(q_0, q_1):			
	$\delta(q_0, 0) = q_4,$	$\delta(q_1, 0) = q_4$	
	$\delta(q_0, 1) = q_1,$	$\delta(q_1, 1) = q_0$	
(q_0, q_4):			
	$\delta(q_0, 0) = q_4,$	$\delta(q_4, 0) = q_0$	
	$\delta(q_0, 1) = q_1,$	$\delta(q_4, 1) = q_2$	✔
(q_1, q_4):			
	$\delta(q_1, 0) = q_4,$	$\delta(q_4, 0) = q_0$	✔
(q_2, q_3):			
	$\delta(q_2, 0) = q_3,$	$\delta(q_3, 0) = q_2$	
	$\delta(q_2, 1) = q_0,$	$\delta(q_3, 1) = q_4$	✔

We now add check marks to distinguish (q_0, q_4) and (q_1, q_4) in preparation for another iteration:

	q_1	q_2	q_3	q_4
q_0		✔	✔	✔
q_1		✔	✔	✔
q_2			✔	✔
q_3				✔

On the next iteration (not shown) q_0 and q_1 remain indistinguishable and can therefore be combined. See Figure 2–50.

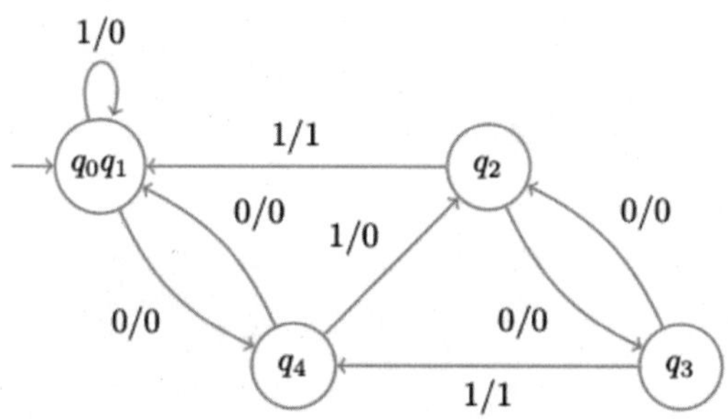

Figure 2–50: Minimal version of Figure 2-49

Key Terms

Mealy machine (aka "finite state transducer") • delayed output

Exercises

1. Draw a Mealy machine that reads ones and zeroes and prints a one every time the substring 1001 is recognized. For example, the input 01001001 yields 00001001.
2. Draw a Mealy machine that performs a circular shift-left operation on strings of a's and b's with an end-of-string marker, so that $abaab\$$ yields the string $baaba$.
3. Draw a Mealy machine that rounds an even binary number up to the nearest odd binary number. This also needs an end marker ($).
4. Draw a Mealy machine that reads bits and prints an even-parity bit after every four bits. You may assume that the input length is a multiple of four. As an example, the input string 10110110 yields 1011101100.
5. Find a Mealy machine for the circuit depicted below.

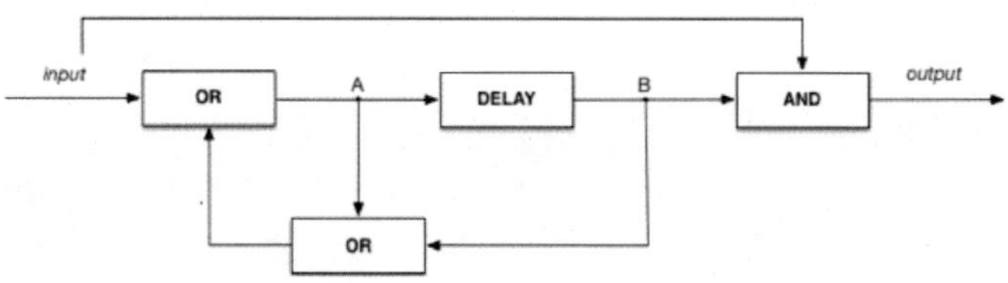

6. Use the procedure described at the end of this section to minimize the following Mealy machine.

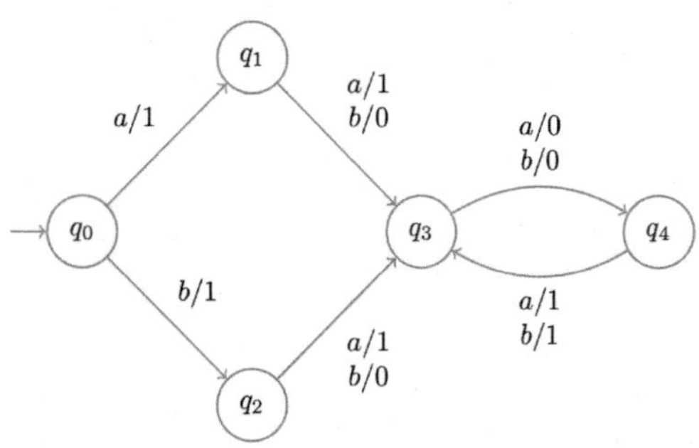

7. Create a machine based on Figure 2–45 to also ignore //-style comments (which ignore everything through the end of the current line) in addition to traditional C-style comments.

Programming Exercises

1. Write a program that implements the comment-ignoring DFA from Exercise 7 above.
2. Implement the machine in Figure 2–46. Use the sample TINY program in this section as input to obtain the expected output.
3. Implement the machine in Exercise 4 above.

Chapter Summary

Deterministic finite automata are useful because they describe so many common scenarios where changes in state occur. They are also straightforward to implement, but can at times be challenging to design correctly. Non-deterministic finite automata are often easier to design than DFAs, but are more difficult to implement. Fortunately, NFAs have an associated DFA that performs the same function, obtained by subset construction. Both types of automata characterize the regular languages. In addition, for every regular language, there is a unique DFA with a minimal number of states.

Finite automata also model processes that convert input into meaningful output, such as filtering out certain text or counting the number of occurrences of substrings. These so-called Mealy machines also have a unique, minimal configuration.

3. Regular Expressions and Grammars

It was as though he found in the chaos of the universe a new pattern, and were attempting clumsily, with anguish of soul, to set it down. I saw a tormented spirit striving for the release of expression.

– *W. Somerset Maugham, The Moon and Sixpence*

"Regular", Terry Shuck Licensed under CC-BY 2.0

Where Are We?

Language Class	Recognized By	Generated By
Regular	Finite Automata	**Regular Grammar** **Regular Expression**
Context-Free	Pushdown Automata	Context-Free Grammar
Recursively Enumerable	Turing Machine	Unrestricted Grammar

Chapter Objectives

- *Use regular expressions to describe regular languages*
- *Convert between regular expressions and finite automata*
- *Use regular grammars to describe regular languages*
- *Convert between regular grammars and finite automata*

Finite automata are useful for implementing acceptors for regular languages, but don't constitute a very compact notation for *describing* regular languages. Consider the language described in English as "one or more a's followed by zero or more c's followed by zero or more b's" (revisit Figure 2–18). Neither English nor an automaton qualify as concise. The notation we used initially in Chapter 2 to introduce that language, however, does qualify as concise: $a^*ac^*b^*$. This is an example of a *regular expression.*

3.1 Regular Expressions

Regular expressions are a convenient notation for representing strings that match simple text patterns[1]. The expression $a^*ac^*b^*$ illustrates two of the four regular expression operations, namely *concatenation* (via juxtaposition) and *Kleene star*. The others appear in the following recursive definition.

Definition 3.1

1. The following are **regular expressions**:
   a) ϕ, representing the empty language/set, $\{\}$
   b) λ, representing the one-string language/set, $\{\lambda\}$
   c) c, for each character, c in the alphabet Σ, representing the language/set $\{c\}$
2. For regular expressions r, r_1, r_2, the following are also regular expressions:
   a) $r_1 + r_2$ (union[a])
   b) $r_1 r_2$ (concatenation)
   c) r^* (Kleene star)
   d) (r) (grouping)

[a] some authors (and most programming environments) use the notation $r_1 \mid r_2$ for union.

Each regular expression represents a language, which is the set of all strings matching its pattern. Formally, we denote the language associated with a regular expression, r, as $L(r)$, so $L(\phi) = \{\}$, $L(\lambda) = \{\lambda\}$, and $L(c) = \{c\}$ for each symbol c from the alphabet in use. The following also hold:

- $L(r_1 + r_2) = L(r_1) \cup L(r_2)$
- $L(r_1 r_2) = L(r_1) L(r_2)$
- $L(r^*) = (L(r))^*$
- $L((r)) = L(r)$

Definition 3.1 is a constructive definition. For example, we can construct the regular expression $aa^*c^*b^*$ as follows:

1. a, b, and c are regular expressions by rule 1–c
2. a^*, b^*, and c^* are regular expressions by rule 2–c
3. $aa^*c^*b^*$ is a regular expression by rule 2–b (applied three times)

We now construct a regular expression representing the language of a's and b's that contain the substring aba. We don't care what precedes or follows the substring aba, so a regular expression for this language is $(a + b)^*aba(a + b)^*$. The formal construction satisfying Definition 3.1 goes like this:

[1]Regular expressions are an example of a *pattern language*.

1. a and b are regular expressions by rule 1–c
2. aba is a regular expression by rule 2–b (applied twice)
3. $a + b$ is a regular expression by rule 2–a
4. $(a + b)$ is a regular expression by rule 2–d
5. $(a + b)^*$ is a regular expression by rule 2–c
6. $(a + b)^*aba(a + b)^*$ is a regular expression by rule 2–b (applied twice)

Here are some more examples of regular expressions over the alphabet $\Sigma = \{a, b\}$:

Table 3–1: Regular expressions for selected languages over $\Sigma = \{a, b\}$

Regular Language	Regular Expression
Strings ending in ab	$(a + b)^*ab$
Strings that begin and end with different symbols	$a(a + b)^*b + b(a + b)^*a$
String of even length	$((a + b)(a + b))^*$
Strings with an even number of a's	$(ab^*a + b)^*$

The precedence of regular expression operators parallels that of algebra:

1. Grouping is the highest-precedence operator
2. Followed by Kleene star (like exponentiation)
3. Followed by concatenation (like multiplication)
4. Followed by union (like addition)

Note the sub-expression ab^*a in the last example in Table 3–1 above. We know that a's must be introduced two at a time to have an even number, but they don't have to be consecutive, so we insert a b^* between the two a's.

Regular expressions constitute a *pattern language* for text processing.

While we are accustomed to using exponents as a shorthand for repetition, using *variables as exponents* is *not* allowed in regular expressions. For example, a^3 as a shorthand for aaa is a regular expression because aaa is a regular expression, but a^n is *not* a regular expression.

a^k where k is a **constant** is a regular expression, but a^n, where n is a **variable** is *not*!

Regular expressions also observe certain algebraic rules, such as a distributive law, wherein $aa + ab = a(a + b)$ and $b + ba = b(\lambda + a)$. Order matters here, because what looks like multiplication is actually concatenation, so $a(a+b) \neq (a+b)a$. The following table summarizes common rules for rewriting regular expressions.

Table 3–2: An algebra for regular expressions

Rule	Description
$r+s=s+r$	Union is *commutative*
$(r+s)+t=r+(s+t)$	Union is *associative*
$r+r=r$	Union is *idempotent*
$r+\phi=r$	Union's identity element is *phi*
$(rs)t=r(st)$	Concatenation is *associative*
$r\lambda=\lambda r=r$	Concatenation's *identity element* is λ
$(r\phi=\phi r=\phi$	Concatenation's *nullity element* is ϕ
$r(s+t)=rs+rt$	Concatenation over union is *distributive*
$(r+s)t=rt+rs$	Concatenation over union is *distributive*
$(r^*)^*=r^*$	Kleene star is *idempotent*
$r^*r^*=r^*$	
$(r+s)^*=(r^*s^*)^*=(s^*r^*)^*$	
$L(r^*)\subseteq L(s^*)\rightarrow r^*s^*=s^*$	
$L(r)\subseteq L(s^*)\rightarrow (r+s)^*=s^*$	

We can conclude, for example, that $(r+\lambda)^*=r^*$, since $L(\lambda)=\lambda\subseteq L(r^*)$. Likewise, $aa\subseteq L(a^*)$, so $(a+aa)^*=a^*$.

Consider also the expression $(a^*+\phi)^*+aa$. Since the empty set adds nothing to the first term, we have $(a^*+\phi)^*=(a^*)^*=a^*$, so the original expression can be written as a^*+aa. However, $(aa)\subseteq L(a^*)$, so the aa is superfluous, making the original expression equivalent to a^*.

Regular expressions are useful for defining patterns for language tokens, as discussed in Example 2–29. An alphabetic identifier, for instance, matches the pattern `<letter><letter>*`, where here we let `<letter>` represent any single letter in our alphabet. Variable names consisting of more than one symbol appear in angle-brackets for clarity. A shorthand notation used by editors and UNIX-like text utilities render this expression using a square-bracket-delimited range: `[A-Za-z][A-Za-z]*`.

Bracket-delimited groups are called *character classes*, representing one of a set of acceptable characters. The class `[aeiou]`, for example, represents any single vowel character. Using the dash specifies a range of characters, so `[A-Za-z]` represents any letter. The plus sign, when used as an *exponent*, means "one or more", so we could just write `[A-Za-z]`$^+$ instead of `[A-Za-z][A-Za-z]*`. A decimal integer in the C language, which must not begin with a zero, matches the pattern `[1–9][0–9]*`. These special abbreviations are not part of the official definition of regular expressions as we are presenting them, but are extensions used for convenience by language designers and text editing programs.

Key Terms

regular expression • pattern language • character class

Exercises

1. Write a regular expression for the language of strings over $\Sigma = \{a, b\}$ containing exactly two a's.
2. Write a regular expression for the language of strings over $\Sigma = \{0, 1\}$ that have an odd number of 1's.
3. Write a regular expression for the language of strings over $\Sigma = \{a, b\}$ whose length is a multiple of 3.
4. Write a regular expression for the language with strings of the form $a^n b^m$ where m and n have the same parity (i.e., the numbers of a's and b's are either both even or both odd simultaneously).
5. Write a regular expression over $\Sigma = \{0, 1\}$ where the bit strings represent 0 or a positive, even number. Leading zeroes are not allowed, except for the number 0 itself.
6. Write a regular expression for the language of strings over $\Sigma = \{0, 1\}$ where the symbols alternate between 0 and 1 (e.g., 0, 1, ... 0101,01010,1010,10101).
7. Why isn't $a^n b^n$ a regular expression? (Think about how you would go about defining a DFA that accepts the language $a^n b^n$? Why can't you do it?)

3.2 Equivalence of Regular Expressions and Regular Languages

Not all languages lend themselves to easy discovery of a regular expression to describe them. Consider, for example, the language consisting of an even number of both a's and b's. We know that each letter must occur in pairs, but not necessarily consecutive pairs. In addition, we can introduce the a's and b's independently–having no a's and/or b's is valid, since zero is an even number. The answer, shown below, is not easy to come up with by ad hoc means:

$$(aa + bb + (ab + ba)(aa + bb)^*(ab + ba))^*$$

It is especially difficult to come up with a regular expression for the complement language of a given regular expression. Regular expressions describe what is *in* a language, not what is outside of it. Fortunately, we can convert any finite automaton to a regular expression, and vice-versa.

To show that regular expressions represent the class of regular languages, we demonstrate the following[2]:

1. For every regular expression, there is a NFA (and therefore a DFA) that accepts the same language.
2. For every finite automaton (DFA or NFA) there is a regular expression that describes the language the automaton accepts.

From Regular Expression to NFA

To show that there is an automaton for every regular expression, it is sufficient to exhibit an automaton for each component of Definition 3.1. The NFAs in Figure 3–1 below represent the base cases.

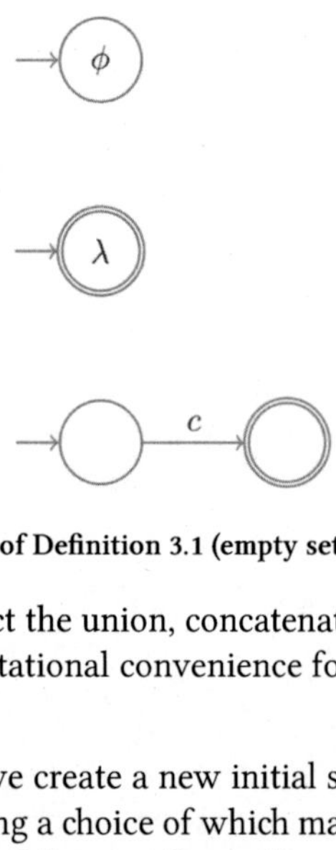

Figure 3–1: NFAs for Part 1 of Definition 3.1 (empty set, empty string, single letter)

We now need to show how to construct the union, concatenation, and Kleene star of arbitrary NFAs. The grouping operation is merely a notational convenience for regular expressions and is not part of the proof.

To construct the *union* of two NFAs, we create a new initial state and use lambda expressions to the start states of each of the NFAs, allowing a choice of which machine to use. If x_1 and y_1 are the initial states of the two NFAs in question, the following figure illustrates the process.

[2]This is Kleene's Theorem, after Stephen C. Kleene.

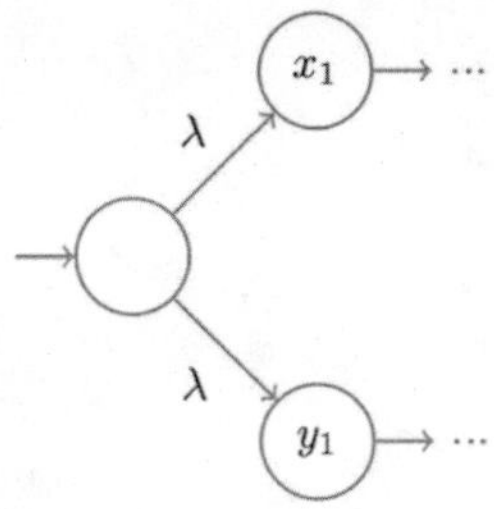

Figure 3–2: NFA fragment representing the union of two other NFAs

To construct the *concatenation* of two NFAs, we place a lambda transition from *each accepting state* of the first machine to the *start state* of the second, and make those final states in the first machine non-final. Of course, the start state of the second machine is also no longer a start state. See Figure 3-3, which assumes, without loss of generality, that x_i and x_j are the final states of the first machine and y_1 is the start state of the second.

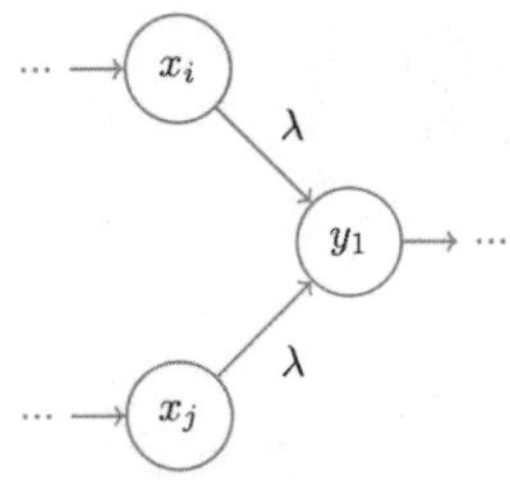

Figure 3–3: NFA fragment representing the concatenation of two other NFAs

To construct the *Kleene star* of a NFA involves two considerations:

1. If the empty string is not already accepted, we must make it accepted.
2. We must allow arbitrary “restarts” of the NFA.

If the initial state of a NFA is also an accepting state, consideration 1 above is already fulfilled. If not, we can make it accept only if there are *no incoming edges* from other states. Otherwise, making the start state accepting would change the language of the machine. In such a case we create a new, accepting start state followed by a lambda transition to the original start state:

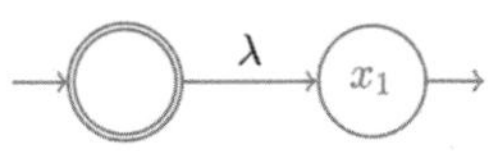

Figure 3–4: Accepting the empty string

To allow arbitrary repetition of a machine, we add lambda transitions from each accepting state to the original initial state, since repetition is just concatenation with self:

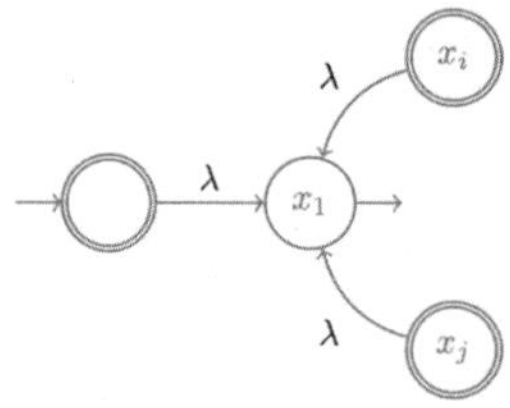

Figure 3–5: Allowing repetition

This constitutes a constructive proof of the first part of Kleene's Theorem. Let's see it in action by constructing a NFA for the regular expression $(ab^*a + b)^*$. We begin with a NFA accepting the singleton string a:

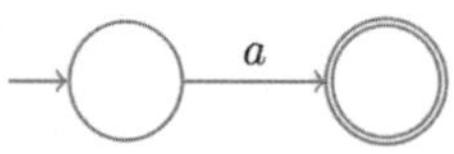

Figure 3–6: NFA accepting a

Next, we allow zero or more b's (we are omitting a lambda transition to a separate machine here for simplicity):

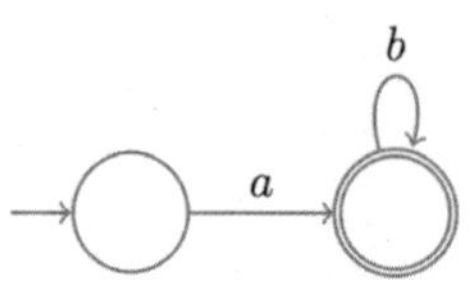

Figure 3–7: Concatenating b^* after a

Now we concatenate another a:

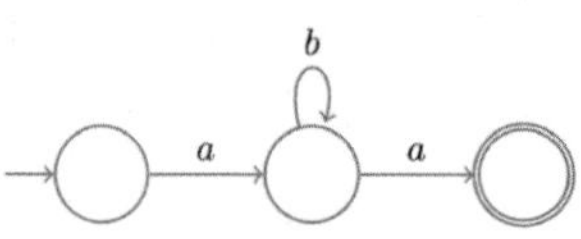

Figure 3–8: Concatenating a

Next, we allow the alternative of just accepting b:

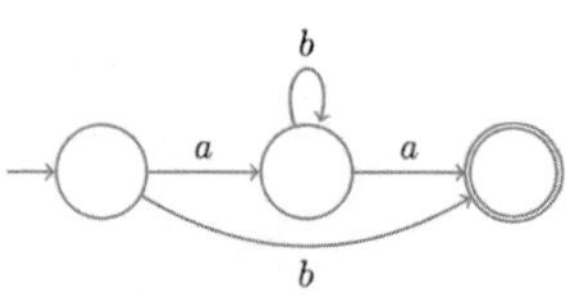

Figure 3–9: Allowing a choice of b

Finally, in Figure 3-10 we form the Kleene star of the machine in Figure 3-9. Since it does not accept the empty string, but also does not have any incoming edges into the start state, we can make the start state accept, as well as add a lambda transition from the original final state back to the original start state for repetition:

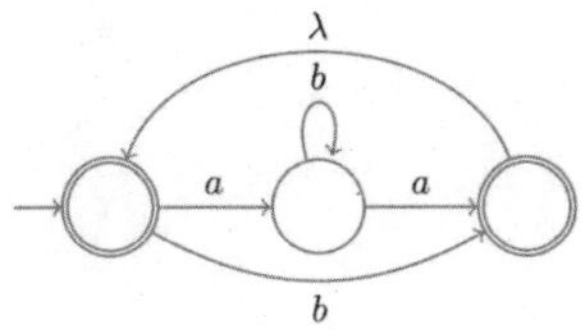

Figure 3–10: The Kleene star of Figure 3–9

From NFA to Regular Expression

To prove the second part of Kleene's Theorem, we need to start with a finite automaton and obtain a regular expression that accepts the same language. The process involves removing states and adding edges so that we end with a two-state machine of the following form (*Note*: expressions $expr_1$ and $expr_3$ are optional):

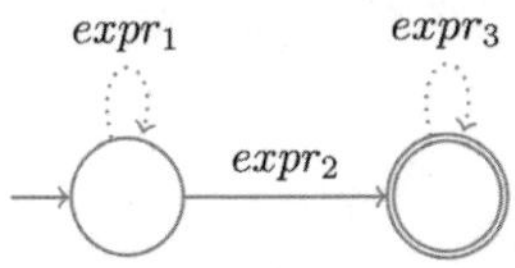

Figure 3–11: Final form of the FA-to-Regular Expression conversion process

The regular expression of such a machine is $(expr_1)^*\,(expr_2)\,(expr_3)^*$. For this process to work, we need to start with a machine with the following properties:

1. There are *no edges entering the initial state* from any other state (self-loops are okay)
2. There are *no edges leaving any final state* to any other state (self-loops are okay)
3. There is *only one final state*, which is not the initial state
4. There is *no jail* (remove it if needed, since it does not contribute to the language)

If the machine is not in this form, add an artificial start and/or a final state with suitable lambda transitions to obtain this form, and delete any jail state.

Example 3–1

To illustrate the conversion process, we convert to a regular expression the machine in Figure 2–1, repeated below, which accepts strings ending with *b*.

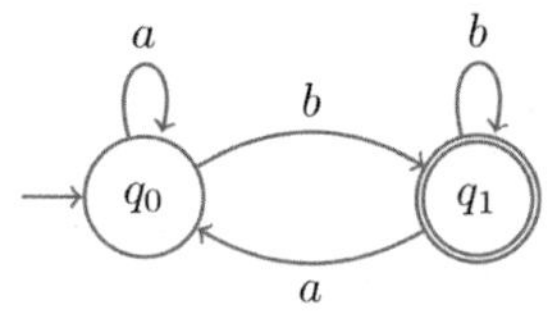

This machine has only one final state, but violates both conditions 1 and 2 above. To put the machine in proper form, we will add new start and final states without changing the language accepted:

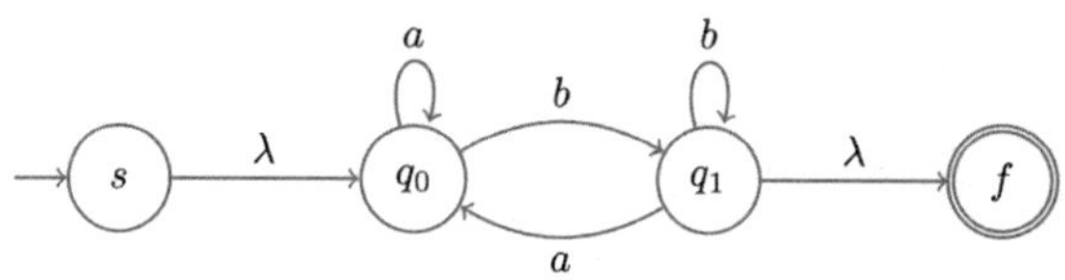

Figure 3–12: Figure 2–1 in the proper form for conversion to a regular expression

We now "eliminate" states q_0 and q_1 while adding edges with strings that represent the paths lost by removing the states[3].

To eliminate state q_0, we need to account for all incoming and outgoing paths *through* q_0 (self-loops don't matter). There are two incoming paths and one outgoing path for q_0, so there are two ways of passing through state q_0, as shown in the following table.

Table 3–3: Paths through state q_1

Path	Equivalent Expression
$s \to q_0 \to q_1$	a^*b
$q_1 \to q_0 \to q_1$	aa^*b

The first path, $s \to q_0 \to q_1$, concatenates λ with zero or more a's followed by a b. We will therefore draw an edge from s to $q1$ containing a^*b, bypassing q_0. For the second path, we will add aa^*b to the self-loop on q_1 (don't omit the original b!). The result of eliminating state q_0 appears in the diagram below.

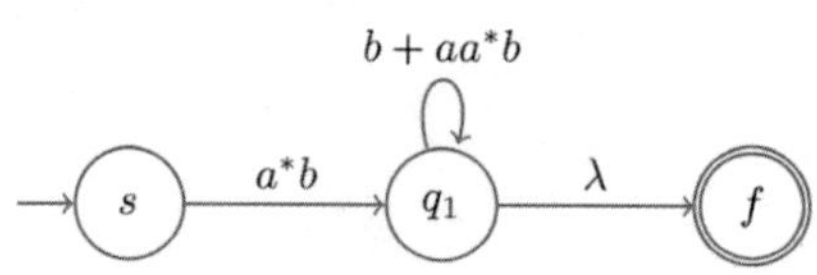

Figure 3–13: Eliminating state q_0

[3]The order of elimination is not important, but it is common that the regular expressions obtained using different orders of elimination are not identical symbol-for-symbol. This is not a problem; the languages the expressions represent will be the same. Eliminating states with the fewest number of through-paths first usually results in the simplest regular expression.

This is no longer a NFA, since we no longer have single characters on the edges. We call such a diagram a *generalized transition graph* (GTG)[4].

We can now eliminate state q_1, since there is only path remaining, giving the regular expression $a^*b(b + aa^*b)^*$. You can see that any string satisfying this regular expression ends with a b. This expression is equivalent to the simpler $(a + b)^*b$.

The number of paths through a state is the number of incoming edges times the number of outgoing edges (ignoring self-loops).

Example 3–2

We now find a regular expression for the *complement* of the language $(a+b)^*aba(a+b)^*$. A DFA for the complement is in Figure 2–7, but it is not in the proper form for conversion to a regular expression. We first remove the jail state and add replacement start and final states (see below).

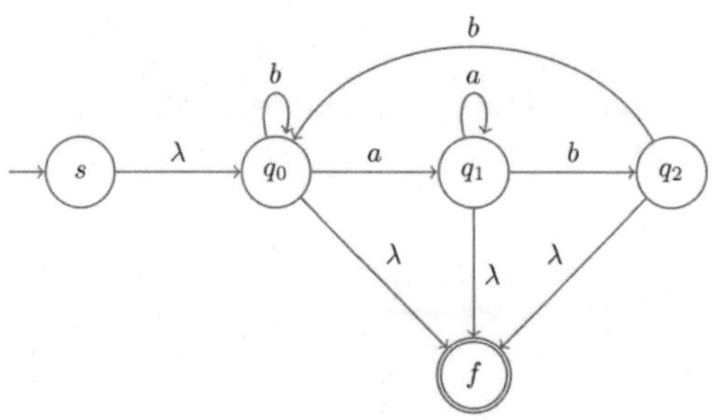

Figure 3–14: Figure 2–7 ready for conversion to a regular expression

State q_0 has two incoming and two outgoing edges, so there are four paths to replace. See the table below.

Table 3–4: Paths through state q_0

Path	Equivalent Expression
$s \to q_0 \to q_1$	b^*a
$s \to q_0 \to f$	b^*
$q_2 \to q_0 \to q_1$	bb^*a
$q_2 \to q_0 \to f$	bb^*

[4]GTGs are sometimes called "expression graphs".

We now add these four new paths and eliminate state q_0:

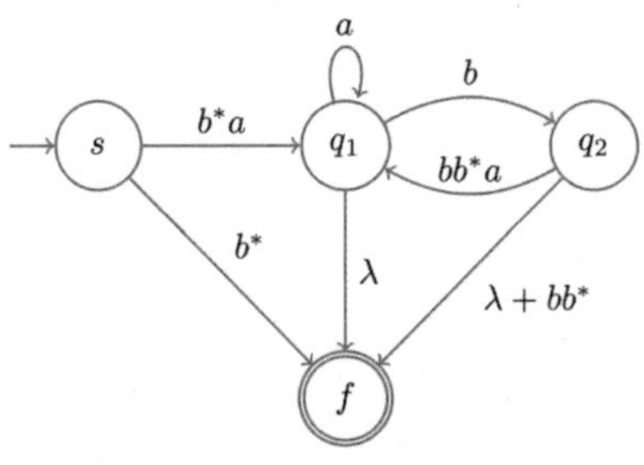

Figure 3-15: GTG after removing q_0

q_2 has only one incoming edge, so let's remove it next[5]. Table 3–5 has the paths to be added.

Table 3–5: Paths through state q_2

Path	Equivalent Expression
$q_1 \to q_2 \to q_1$	bbb^*a
$q_1 \to q_2 \to f$	$b(\lambda + bb^*)$

The next figure updates the graph by removing q_2 and adding the required edges.

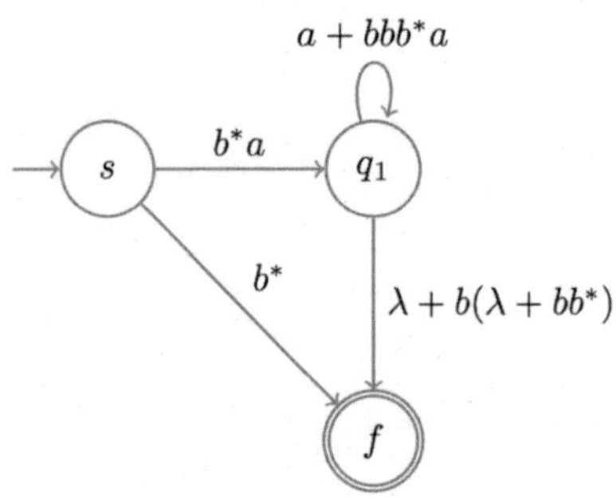

Figure 3–16: GTG after removing q_2

We now see two alternate paths from s to f, from which we form a union for the final answer:

$$b^*a(a + bbb^*a)^*(\lambda + b(\lambda + bb^*)) + b^*$$

---

[5]Again, order is not important, but with experience you can see which order might result in less tedium. And remember, although the regular expression may change if you change the order of elimination, the result is equivalent in that it represents the same language.

Key Terms

Kleene's Theorem • generalized transition graph (GTG) / expression graph

Exercises

1. Find a NFA for the regular expression $a^*(b(a + aaa))^*$.
2. Using the state-bypass and elimination technique, convert the FA in Figure 2–17 (repeated below) to a regular expression.

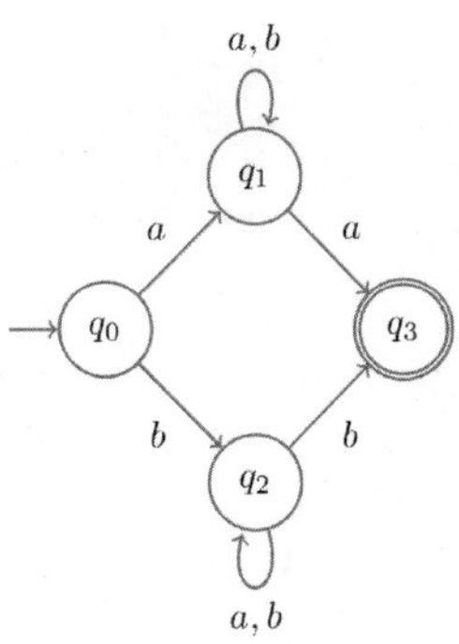

3. Using the state-bypass and elimination technique, convert the FA in Figure 2–22 (repeated below) to a regular expression.

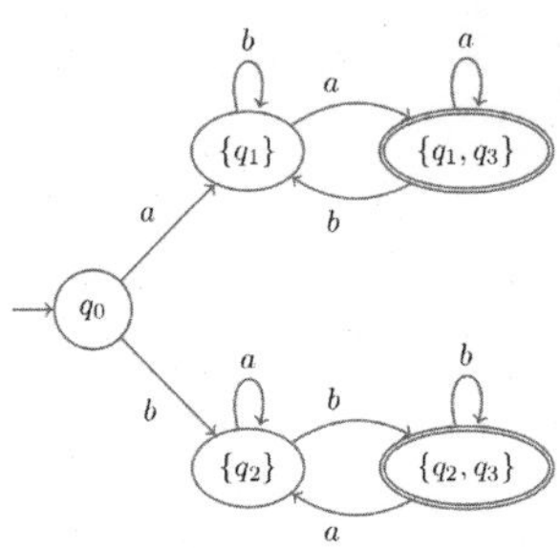

4. Find a NFA that accepts the concatenation of the language accepted by the machine in Figure 2–6 followed by the machine in Figure 2–4. (Machines repeated below.)

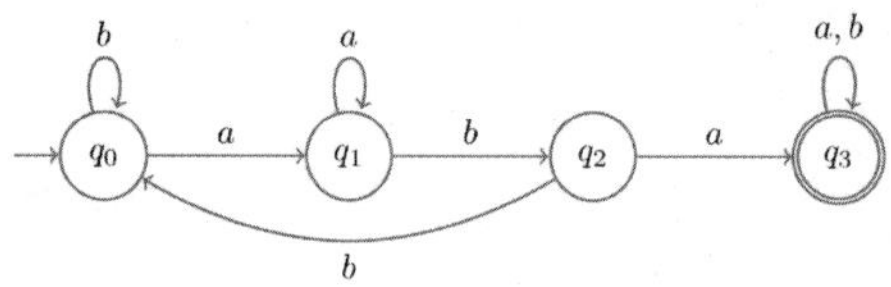

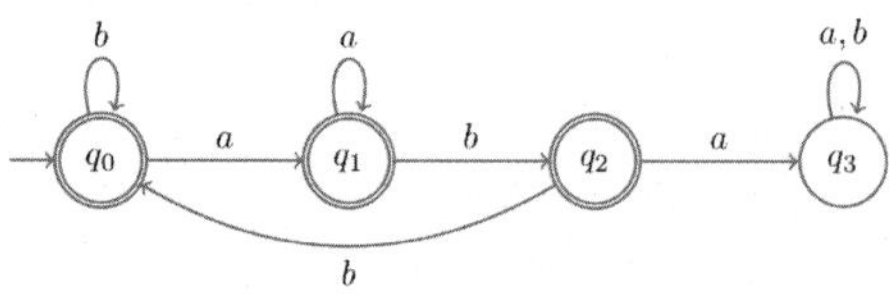

5. Find a NFA that accepts the Kleene star of the language accepted by the machine in Figure 2–8.
6. Using the state-bypass and elimination technique, convert the machine in the previous exercise (#5) to a regular expression.
7. Describe a procedure for determining whether two *regular expressions* describe the same language.

3.3 Regular Grammars

We saw in Chapter 1 that formal languages generate strings in a language. Grammars that generate regular languages have a special form, which naturally corresponds to a finite automaton. To illustrate, the NFA from Example 2–9, which accepts strings of ones and zeroes with a one in the third-to-last place, appears below with the states renamed for convenience.

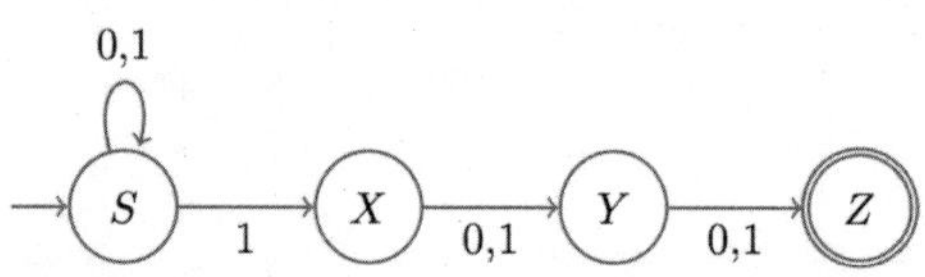

Figure 3–17: NFA accepting strings with 1 in the third-to-last place

The state names serve as variable names in the grammar. From state S we can loop on S with either character, or we can move to state X with a 1. The following rewrite rule expresses this in formal grammar terms: $S \rightarrow 0S \mid 1S \mid X$. We can similarly express the transitions from X and Y. Since Z is a final state, we allow it to be replaced by the empty string, meaning that the generation of symbols may terminate. The complete grammar appears below.

$$
\begin{aligned}
S &\rightarrow 0S \mid 1S \mid 1X \\
X &\rightarrow 0Y \mid 1Y \\
Y &\rightarrow 0Z \mid 1Z \\
Z &\rightarrow \lambda
\end{aligned}
$$

Observe the clear correspondence to the NFA. The from-state appears alone on the left, and the to-states appear on the right, preceded by the symbol from the appropriate edge. This is an example of a **right-linear grammar**. It is "linear" because only *one* variable appears in each right-hand side. It is "*right*-linear" because the variable appears *last* (i.e., on the *right end*) of each replacement rule. We can derive a sample string as follows:

$$S \Rightarrow 0S \Rightarrow 01X \Rightarrow 010Y \Rightarrow 0100Z \Rightarrow 0100$$

Definition 3.2

A **right-linear grammar** is a formal grammar where there is at most one variable in any right-hand side rule, and if present, it is the rightmost symbol of the rule.

More formally, rules in a right-linear grammar are of form $V \rightarrow sX$ or $V \rightarrow s$, where V and X are variables and s is any string in Σ^* (i.e., s contains no variables).

Example 3–3

Suppose now that we want to generate strings of ones and zeroes representing binary numbers congruent to 1 mod 3. We begin by renaming the states in Example 2–6 and making the state where the remainder modulo 3 is 1 the accepting state (Z for 0, O for 1, T for 2, representing the remainders):

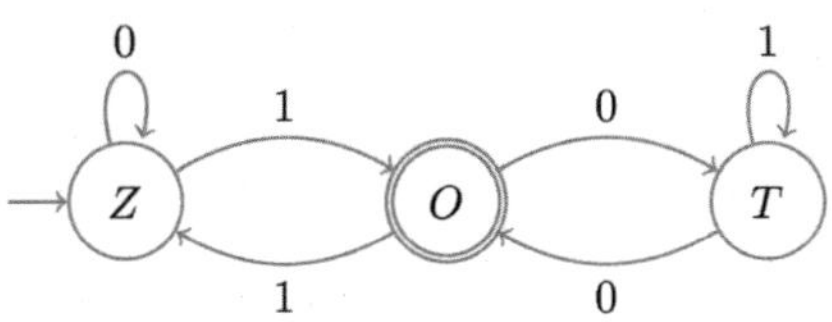

Figure 3–18: DFA accepting binary numbers congruent to 1 mod 3

As before, we echo the paths with grammar rules, and allow the accepting state to become the empty string:

$$
\begin{aligned}
Z &\to 0Z\ |1O \\
O &\to 0T \mid 1Z \mid \lambda \\
T &\to 0O \mid 1T
\end{aligned}
$$

Any string we generate will have the appropriate remainder:

$$Z \to\Rightarrow 1O \Rightarrow 10T \Rightarrow 100O \Rightarrow 100$$

This derivation yields $100_2 = 4_{10} \equiv 1 \bmod 3$.

We summarize the process for converting a finite automaton to a right-linear grammar below:

1. Remove any jail states (they do not contribute to the language)
2. For each transition, form a grammar rule as follows:
   a. Place the variable representing the from-state on the left-hand side of the rule
   b. Form a string for the right-hand side of the rule by concatenating the character (or string, in the case of a GTG) on the transition followed by the variable representing the to-state
3. For each accepting state, X:
   a. Add the rule $X \to \lambda$

Converting a right-linear grammar to a transition graph is the reverse of this process. To illustrate, we convert the following regular grammar to a DFA.

$$
\begin{aligned}
S &\to aX \mid bY \mid \lambda \\
X &\to aS \mid bZ \\
Y &\to aZ \mid bS \\
Z &\to aY \mid bX
\end{aligned}
$$

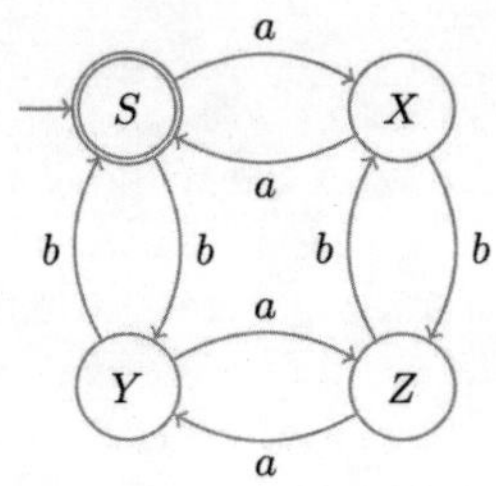

Figure 3–19: DFA for the grammar above

A right-linear grammar represents a regular language even if there is more than one character preceding the variable. The following grammar represents the same language as the grammar above.

$$S \to abT \mid baT \mid aaS \mid bbS \mid \lambda$$
$$T \to abS \mid baS \mid aaT \mid bbT$$

Instead of a DFA, we get a GTG from this grammar:

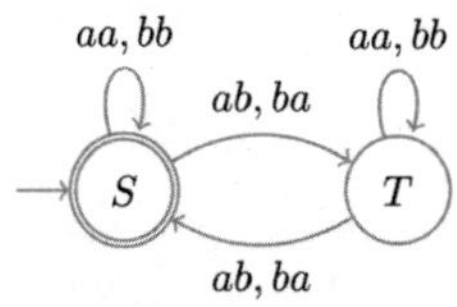

Figure 3–20: DFA for the previous grammar

Can you discern what language these grammars and graphs represent?

If a regular grammar has a rule of the form $V \to s$, where s is any non-empty string in Σ^*, then we create a new accepting state as the target of an edge from V with s as its text, as shown in the following figure.

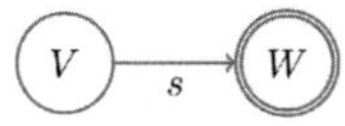

Figure 3–21: Subgraph for the rule $V \to s$

This is because $V \to s$ can be rewritten as

$$V \to sW$$
$$W \to \lambda$$

and the λ can be back-substituted.

Left-Linear Grammars

It is also possible to represent a regular language with a *left-linear* grammar, although the correspondence to an automaton is not as clear at first. Left-linear grammars generate strings *backwards* (right-to-left), whereas right-linear grammars generate strings left-to-right. We extract a left-linear grammar from a finite automaton or GTG by traversing the machine backwards. If there are multiple final states, we introduce a new, single final state with lambda moves from the original final states, as we did when converting automata to regular expressions earlier. The final state then becomes the start state, and vice-versa. The trick is to write the rules backwards as we go.

Example 3–4

In this example, we find a left-linear grammar for the language aab^*a. We first draw a GTG for this language:

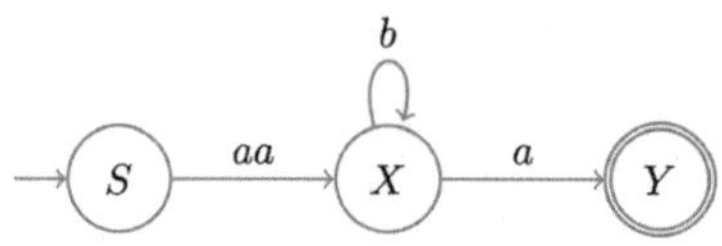

Figure 3-22: GTG for aab^*a

Y will be the start variable in a left-linear grammar for this language, since we want to traverse the machine backwards. Y is "preceded" by an a from state X, which we express by the following "backward" rule:

$$Y \to Xa$$

Now we notice that X has two antecedents:

$$X \to Xb \mid Saa$$

We make S accepting since we end there (the initial state becomes the final state when traversing backwards). The complete grammar is, therefore:

$$\begin{aligned} &Y \to Xa \\ &X \to Xb \mid Saa \\ &S \to \lambda \end{aligned}$$

which simplifies to

$$Y \to Xa$$
$$X \to Xb \mid aa$$

To verify that we get a string that matches aab^*a, we derive $aabbba$:

$$Y \Rightarrow Xa \Rightarrow Xba \Rightarrow Xbba \Rightarrow Xbbba \Rightarrow aabbba$$

Example 3–5

The following NFA accepts strings from $(a+b+c)^*$ that are missing at least one of the letters of its alphabet, strings such as $\lambda, a, b, c, ab, ac, bc, abbab, caccac$ and cbb. No string is to contain all three letters simultaneously.

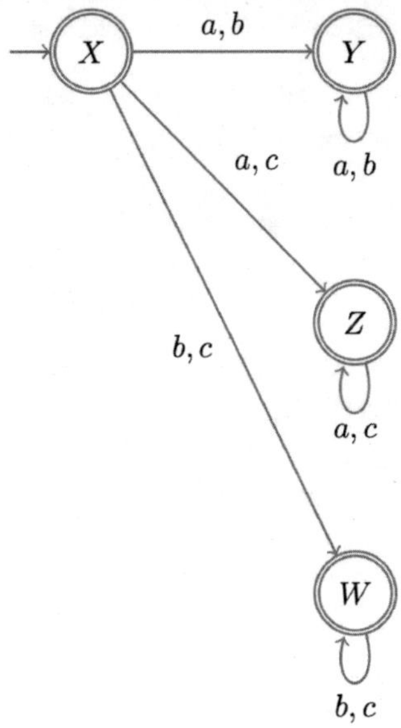

Figure 3–23: Accepts strings in $(a+b+c)^*$ missing at least one of the letters

We first modify the machine to have only one final state, as shown below.

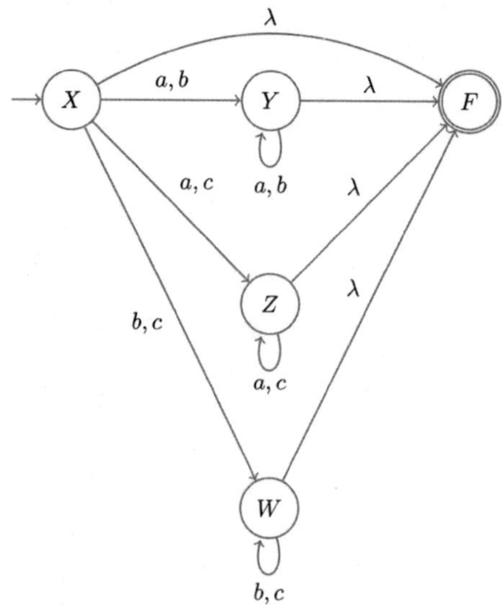

Figure 3–24: Adding a unique final state

We now start in the final state, F, creating left-linear rules, and work our way back to the start state, X. Since F comes from all the other states with an empty string, we have

$$F \rightarrow X \mid Y \mid Z \mid W$$

State Y comes from itself and X with an a or a b, giving the following rules:

$$Y \rightarrow Ya \mid Yb \mid Xa \mid Xb$$

States Z and W behave similarly, so the complete grammar is:

$$\begin{aligned}
F &\rightarrow \lambda \mid Y \mid Z \mid W \\
Y &\rightarrow Ya \mid Yb \mid Xa \mid Xb \\
Z &\rightarrow Za \mid Zc \mid Xa \mid Xc \\
W &\rightarrow Wb \mid Wc \mid Xb \mid Xc \\
X &\rightarrow \lambda
\end{aligned}$$

We can now substitute for X throughout giving the final, simplified grammar:

$$
\begin{aligned}
F &\to \lambda \mid Y \mid Z \mid W \\
Y &\to Ya \mid Yb \mid a \mid b \\
Z &\to Za \mid Zc \mid a \mid c \\
W &\to Wb \mid Wc \mid b \mid c
\end{aligned}
$$

Example 3–6

It is possible to start with a left-linear grammar and obtain a right-linear grammar representing the same language. We simply build the machine backwards from the left-linear grammar, and then extract a right linear grammar from the machine, traversing it forward. Consider the following left-linear grammar:

$$
\begin{aligned}
C &\to Bb \\
B &\to Aa \\
A &\to Bab \mid \lambda
\end{aligned}
$$

We see that C is preceded by a b from state B, B is preceded by an a from state A, and A is preceded from B with the string ab. Since A goes to λ, it is a stopping point going backwards, which makes it the start state going forward. This gives the following machine.

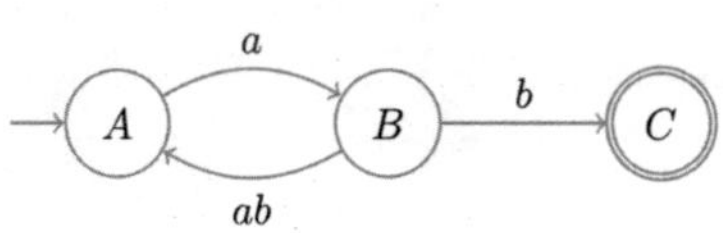

Figure 3–25: GTG for the grammar above

At this point we make A the start variable to obtain an equivalent right-linear grammar:

$$
\begin{aligned}
A &\to aB \\
B &\to abA \mid bC \\
C &\to \lambda
\end{aligned}
$$

which simplifies to

$$A \to aB$$
$$B \to abA \mid b$$

Mixing right and left-linear rules in the same grammar does *not* constitute a regular grammar. Rules must be uniformly left-linear or uniformly right-linear. The following grammar, although linear, is not a regular grammar (it has both a left-linear rule and a right-linear rule):

$$S \to aB \mid \lambda$$
$$B \to Sb$$

Consider the following sample derivation using this grammar:

$$S \Rightarrow aB \Rightarrow aSb \Rightarrow aaBb \Rightarrow aaSbb \Rightarrow aabb$$

This grammar generates the language $a^n b^n, n \geq 0$, which, as we will see in the next chapter, is *not* a regular language. (Try to draw a DFA for it to see why.)

Mixing right and left-linear rules in the same grammar does *not* constitute a regular grammar!

Key Terms

regular grammar • linear grammar • right-linear grammar • left-linear grammar

Exercises

1. Find a Right-linear grammar for the machine in Exercise 2.1.4 (repeated below).

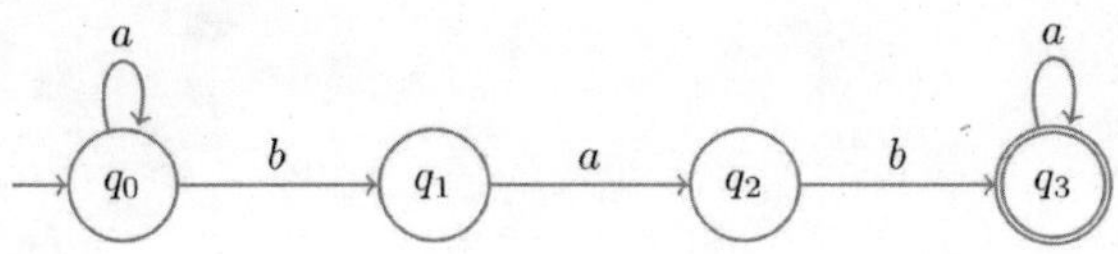

2. Find a Left-linear grammar for the machine in Exercise 1 above.
3. Find a Right-linear grammar for the machine in Figure 2–16 (repeated below).

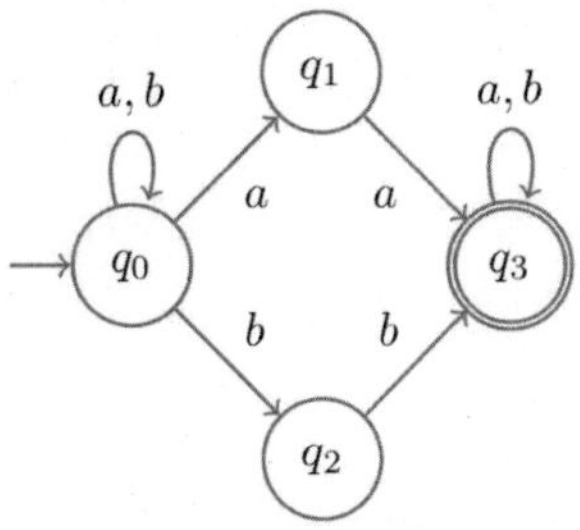

4. Find a Left-linear grammar for the machine in Exercise 3 above.
5. Find a Left-linear grammar that generates the same language as the following Right-linear grammar.

$$
\begin{aligned}
S &\to bS \mid aX \\
X &\to aX \mid bY \\
Y &\to aX \mid bS \mid \lambda
\end{aligned}
$$

6. Find a Right-linear grammar for the language of the following Left-linear grammar.

$$
\begin{aligned}
S &\to Sa \mid Sb \mid Za \\
Z &\to Yb \\
Y &\to Ya \mid Xa \\
X &\to Xb \mid Zb \mid \lambda
\end{aligned}
$$

7. Find a right-linear grammar that generates the language represented by the regular expression $(a+b)b(aa)^*(a+b)^*$.

Chapter Summary

Regular expressions constitute a convenient pattern language for expressing text patterns commonly found in documents and source code. Sometimes it is easier to come up with a finite automaton for a regular language, and sometimes it is easier to find a regular expression. Being able to convert from one form to the other provides a robust toolbox for programming text-processing applications.

Regular grammars, which are special cases of more general grammars that we will see later, are useful for generating strings in a regular language. They are also easy to program, as we will see when we study context-free grammars in Chapter 6.

4. Properties of Regular Languages

I'm just a regular person who believes life is simple, and I like a simple life.

– Manny Pacquiao

"Just Pump", by Katrien Berckmoes Licensed under CC-BY 2.0

Where Are We?

Language Class	Recognized By	Generated By
Regular	Finite Automata	Regular Grammar Regular Expression
Context-Free	Pushdown Automata	Context-Free Grammar
Recursively Enumerable	Turing Machine	Unrestricted Grammar

Chapter Objectives

- *Determine which operations preserve the "regular-ness" of a language*
- *Develop algorithms for answering certain questions about regular languages*
- *Show that some languages are not regular*

In section 3.2 we showed that the union, concatenation, and Kleene star of regular languages are also regular by exhibiting NFAs that accept them. The term describing the property of operators "staying within the same class of languages" is called **closure**; just as integers are **closed** under addition, subtraction, and multiplication, we say that regular languages are closed under union, concatenation, and Kleene star.

A set is closed under an operation if the result of the operation remains in that set.

In this chapter, we take a deeper look at closure and other properties of regular languages, develop algorithms for making decisions about regular languages, and finally, we introduce formal languages that are *not* regular.

4.1 Closure Properties

We showed in Chapter 2 that regular languages are closed under complementation by inverting the acceptability of each state in a DFA that accepts a language. Having thus found an automaton that accepts the complement of the original language, we conclude that the complement of any regular language is regular.

We can show that reversing the strings in a regular language results in another regular language by "reversing" its machine. To reverse a finite automaton, we do the following:

- Remove any jail state[1]
- If there is more than one accepting state, create a new, unique, accepting state, reachable from what were the accepting states, by lambda transitions
- Make the original start state the new, unique, accepting state
- Make the original (now unique) accepting state the new start state
- Reverse all edges (and their *strings*, if the diagram is a GTG)

Example 4–1

To illustrate, we reverse the machine in Figure 2-22, which accepts the language $aa^*c^*b^*$.

[1]Jail states become unreachable when arrows are reversed.

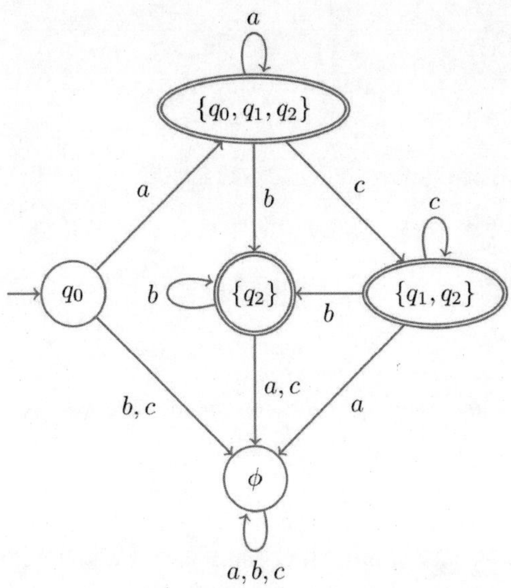

Copy of Figure 2–22

Since there are multiple accepting states, we begin by creating a unique accepting state with incoming lambda transitions. We also remove the jail:

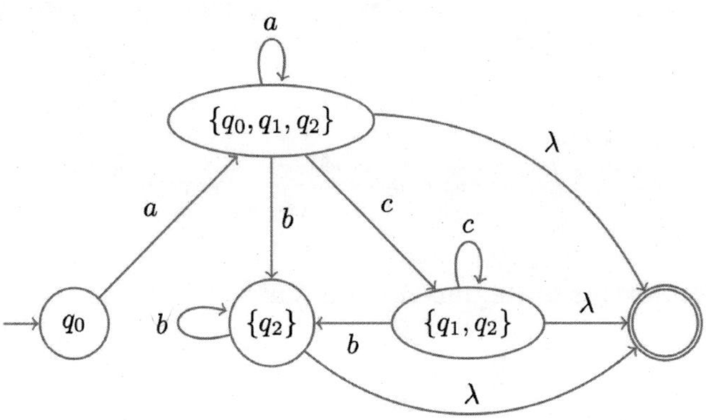

Figure 4–1: Modifying Figure 2–22

We now reverse the arrows as well as the roles of the initial and final state to obtain the following NFA for the reverse language:

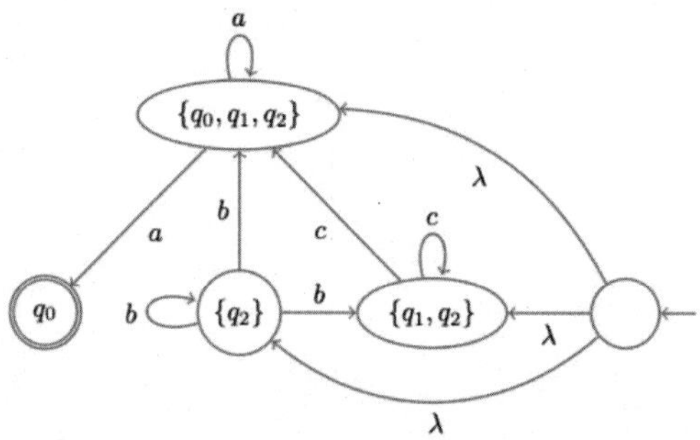

Figure 4–2: The reversal of the previous figure

This machine accepts $b^*c^*a^*a$, a regular expression for the reversal of the original language.

Since languages are *sets*, we consider other set operations for closure. It is easy to show that closure under union and complement are sufficient to guarantee closure under intersection and difference by the following set identities[2].

Operation	Notation	Equivalent Expression
Intersection	$R \cap S$	$\overline{\overline{R} \cup \overline{S}}$
Difference	$R - S$	$R \cap \overline{S}$
Symmetric Difference	$R \ominus S$	$(R - S) \cup (S - R)$

Since regular languages are closed by the operations used on the right-hand sides above, the sets on the left are also regular by construction.

We conclude that the class of regular languages is closed under the following operations.

Closure Properties of Regular Languages

- Union
- Concatenation
- Kleene Star
- Complement
- Reversal
- Intersection
- Set Difference
- Set Symmetric Difference

[2]These can be verified using Venn diagrams.

It is important to understand that arbitrary subsets of regular languages are generally not regular. As a counterexample (to be proven in Section 4.3), the language $\{a^n b^n \mid n \geq 0\}$ is a subset of a^*b^*, but it is *not* regular.

An arbitrary subset of a regular language is not necessarily regular.

Computing Set Operations

We showed previously how to compute the union, concatenation, Kleene star, complement, and reversal of regular languages by modifying or combining automata as needed, but how do we construct the *intersection* of two regular languages? The answer is to track the two machines concurrently and then assign accepting states only where both machines accept simultaneously.

The process is similar to the Rabin-Scott subset construction process used in Section 2.2, and gives us the union, intersection, and differences all at once through a *product table.* The product table process uses DFAs (not NFAs) and begins with the start states of the two automata as its composite, initial state, and we simply track where the machines take us simultaneously for each possible input. Since we need to track all possible machine movements, we must start with DFAs.

Example 4–2

Let's form the product table for the two automata below.

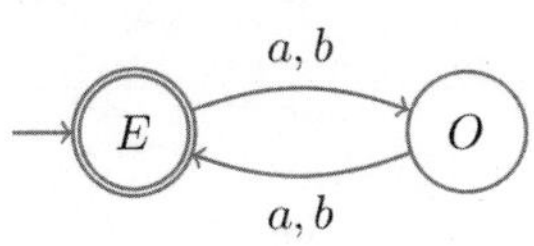

Copy of Figure 1–2

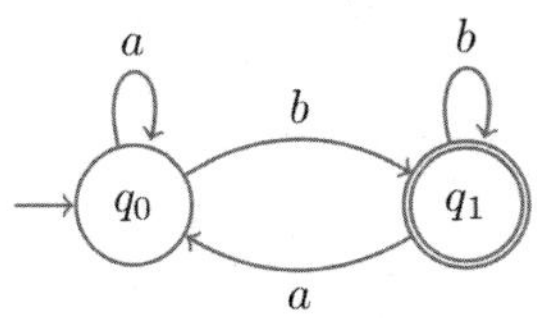

Copy of Figure 2–1

To track the two machines simultaneously, we begin with the composite state $\{E, q_0\}$ as the initial state for the product table:

State	a	b
$\{E, q_0\}$	$\{O, q_0\}$	$\{O, q_1\}$

Table 4–1: Initializing the Product Table

Upon reading any symbol, the first machine switches states, so it first moves from state E to state O. The second machine moves respectively to q_0 or q_1 when reading an a or b. We add the new states just reached and see where the machines move from there:

State	a	b
$\{E, q_0\}$	$\{O, q_0\}$	$\{O, q_1\}$
$\{O, q_0\}$	$\{E, q_0\}$	$\{E, q_1\}$
$\{O, q_1\}$	$\{E, q_0\}$	$\{E, q_1\}$

Table 4–2: After reaching two new states

A new state, $\{E, q_1\}$, has been reached. After following its movements we get the final product table below.

State	a	b
$\{E, q_0\}$	$\{O, q_0\}$	$\{O, q_1\}$
$\{O, q_0\}$	$\{E, q_0\}$	$\{E, q_1\}$
$\{O, q_1\}$	$\{E, q_0\}$	$\{E, q_1\}$
$\{E, q_1\}$	$\{O, q_0\}$	$\{O, q_1\}$

Table 4–3: The final product table

We determine the union, intersection, or the differences by how we assign accepting states to the product machine. If we want the union, then any composite state containing either E or q_1 will be accepting, namely $\{E, q_0\}$, $\{O, q_1\}$, and $\{E, q_1\}$. For the intersection, only $\{E, q_1\}$ accepts.

For the difference of the first minus the second language, we want to simultaneously accept in the first and not in the second, so only $\{E, q_0\}$ accepts for that difference. Conversely, only $\{O, q_1\}$ accepts for the difference of the second minus the first. The symmetric difference is the union of the preceding two, so both $\{E, q_0\}$ and $\{O, q_1\}$ would accept in that case.

The thing to keep in mind is that the structure of the product table/machine is the same for all the set operations; only the acceptability of the resulting composite states differs. A transition graph for the intersection appears below.

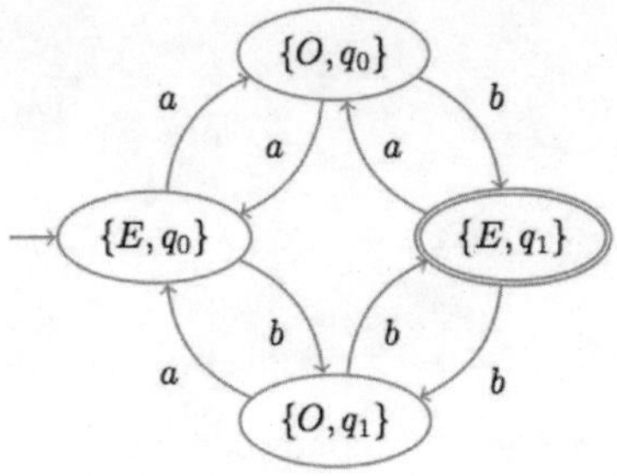

Figure 4–3: The intersection as a transition graph

A **product table** can determine the union, intersection, and differences of any two regular languages.

Example 4–3

Let's find an automaton for the language over $\Sigma = \{0, 1\}$ that accepts strings beginning with a 1 and that do *not* have 00 as a substring. We know the result is a regular language since it is the difference of two regular languages. We begin by finding an automaton for each language (the second automaton below *does* contain a 00, however; see Figure 4-4 and Figure 4-5).

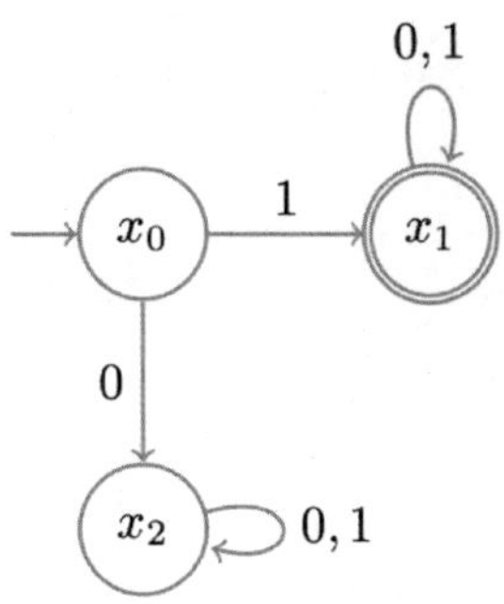

Figure 4–4: Accepts bit strings starting with 1

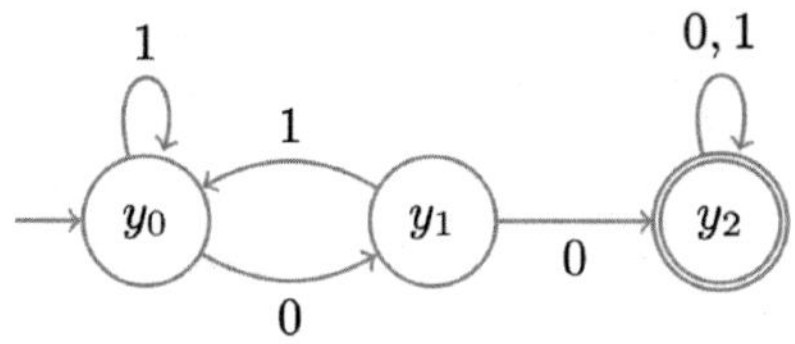

Figure 4–5: Accepts bit strings containing 00

We form the product table by starting with the composite start state and seeing what paths ensue:

State	0	1
$\{x_0, y_0\}$	$\{x_2, y_1\}$	$\{x_1, y_0\}$
$\{x_2, y_1\}$	$\{x_2, y_2\}$	$\{x_2, y_0\}$
$\{x_1, y_0\}$	$\{x_1, y_1\}$	$\{x_1, y_0\}$
$\{x_2, y_2\}$	$\{x_2, y_2\}$	$\{x_2, y_2\}$
$\{x_2, y_0\}$	$\{x_2, y_1\}$	$\{x_2, y_0\}$
$\{x_1, y_1\}$	$\{x_1, y_2\}$	$\{x_1, y_0\}$
$\{x_1, y_2\}$	$\{x_1, y_2\}$	$\{x_1, y_2\}$

Table 4–4: Product table for Figure 4–4 and Figure 4–5

We are interested in the *difference* of the two languages, so we pair *accepting* states in the first with *non-accepting* states in the second, giving the composite accepting states $\{x_1, y_0\}$ and $\{x_1, y_1\}$. The transition graph appears in Figure 4–6 below.

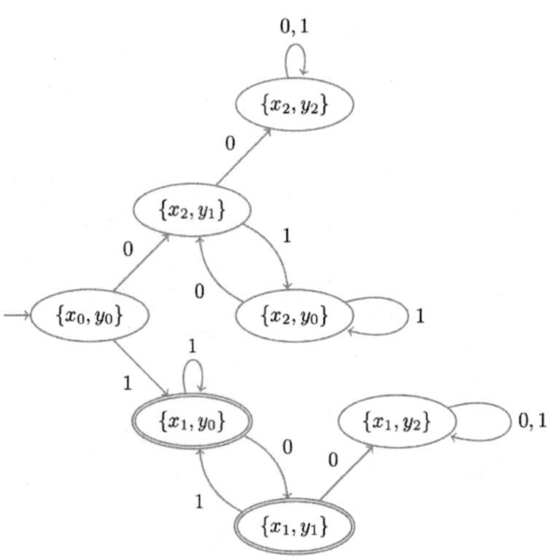

Figure 4–6: Difference of the two languages above

The following table summarizes how to determine the various operations applied to two regular languages from the states in their product table.

Table 4–5: How to determine operations between languages

Operation	Accepting States
$L_1 \cup L_2$	One or both original final states accept
$L_1 \cap L_2$	Both original final states accept
$L_1 - L_2$	L_1 state accepts, L_2 state rejects
$L_2 - L_1$	L_1 state rejects, L_2 state accepts
$L_1 \ominus L_2$	$L_1 - L_2 \cup L_2 - L_1$

Key Terms

closure • product table • composite state

Exercises

1. Find a NFA that accepts the reversal of the language of the machine in Figure 2–9 (repeated below) by reversing the DFA in that figure:

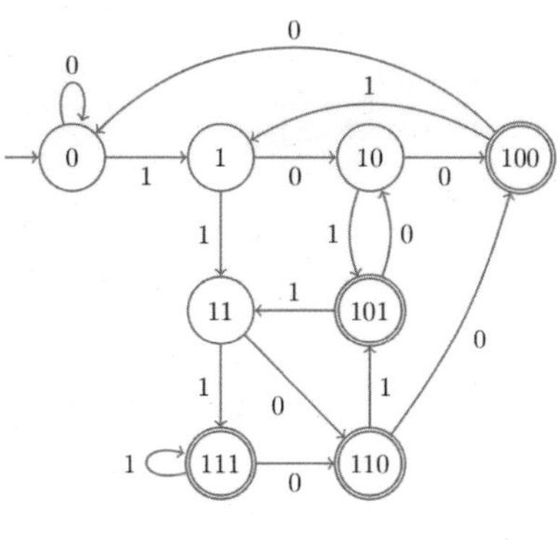

Copy of Figure 2–9

2. Construct the product table for the language $R = aab^*$ and the language S, which contains all strings of a's and b's whose length is a multiple of 3. (Use state names $r_0, r_1, \ldots$ and $s_0, s_1 \ldots$). You will use this in the problems that follow.
3. Which composite states from your product table must accept for $R \cup S$?

4. Which composite states from your product table must accept for $R \cap S$?
5. Which composite states from your product table must accept for $R - S$?
6. Which composite states from your product table must accept for $S - R$
7. Which composite states from your product table must accept for $R \ominus S$?

4.2 Decision Algorithms

We can now establish some procedures that enable us to make certain decisions about regular languages. For example, we might ask the question, "Do two automata recognize the same regular language?" Two languages are equal if they are the same set of strings. We can use results from the previous section to develop a procedure to answer this question.

Two sets are equal if and only if they have the same elements. When the sets are infinite, the best way to establish their equality is to show that any element in one set is also in the other, and vice-versa; that is, $L_1 = L_2$ if and only if $L_1 \subseteq L_2$ *and* $L_2 \subseteq L_1$. Another way to express this idea is that neither set has an element that the other doesn't. The latter statement is equivalent to the mathematical expression $L_1 \ominus L_2 = \phi$. Since we know how to compute the symmetric difference of two automata, this is our algorithm for testing language equality.

Example 4–4

We know that the machines in Figure 2–29 and Figure 2–31 accept the same language since the second is the minimization of the first. To illustrate the procedure testing for equality of languages, let's verify that the symmetric difference of those two machines is empty. The machines appear below with the states of the second machine renamed for convenience.

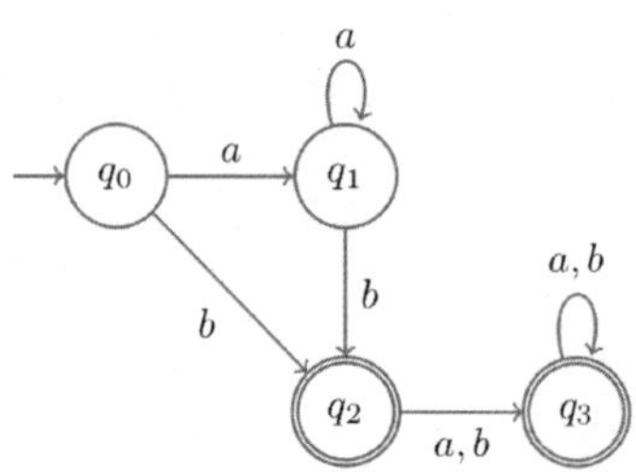

Copy of Figure 2–29

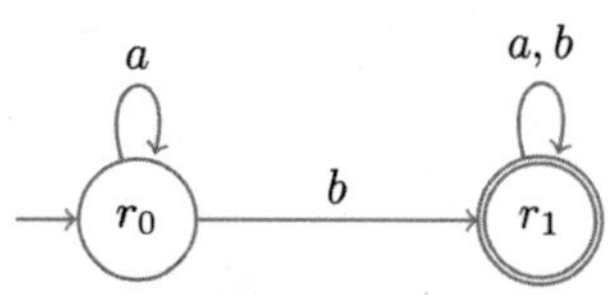

Figure 2–31 with states renamed

Calling the first machine Q and the second R, wef form the product table to determine the symmetric difference, $Q \ominus R$. See the following table.

State	a	b
$\{q_0, r_0\}$	$\{q_1, r_0\}$	$\{q_2, r_1\}$
$\{q_1, r_0\}$	$\{q_1, r_0\}$	$\{q_2, r_1\}$
$\{q_2, r_1\}$	$\{q_3, r_1\}$	$\{q_3, r_1\}$
$\{q_3, r_1\}$	$\{q_3, r_1\}$	$\{q_3, r_1\}$

Table 4–6: Product table for the two automata above

Since $Q \ominus R = (Q - R) \cup (R - Q)$ we look at the differences separately.

A machine representing $Q - R$ would have as accepting states those where Q accepts and R rejects, namely the composite states $\{q_2, r_0\}$ and $\{q_3 r_0\}$. These combinations *do not appear* in the product table; they are not possible when traversing these two machines simultaneously with the same string. We conclude that $Q - R = \phi$.

For $R - Q$, composite states $\{q_0, r_1\}$ and $\{q_1, r_1\}$ represent where R accepts and Q rejects. These states do not appear in the product table either, so we conclude that $Q - R$ is empty, and therefore so is $Q \ominus R$. Neither machine accepts a string that the other doesn't, so they accept the same language.

Example 4–5

As an example of showing two languages unequal, we look at the symmetric difference of the machines in Figure 4–4 and Figure 4–5. The difference $X - Y$ requires that states $\{x_1, y_0\}$ or $\{x_1, y_1\}$ must appear in the product table. Inspecting Table 4–4, we see that, in fact, both composite states are present, so we immediately conclude that $X - Y \neq \phi \Rightarrow X \ominus Y \neq \phi$; the languages are not the same.

In case it crossed your mind, using the state minimization procedure to determine language equality of two finite automata is not always the preferred approach (see Exercise 3.2-7) because it remains to show whether the minimized transition graphs are isomorphic—not always an easy process. Use the *symmetric difference* to determine if two automata accept the same language. If you have regular expressions instead of automata, convert them to automata first and then proceed to compute their symmetric difference by their product table.

Use the **symmetric difference** to determine equality of two regular languages from their automata.

Having established an effective decision procedure to determine whether two regular languages are equal, we now consider the following questions:

- Is one regular language a subset of another?
- Is the language of a given DFA empty?
- Is the language of a given DFA infinite?

To show that one regular language is a subset of another, $X \subseteq Y$, say, it is sufficient to show that X does not have any strings that Y doesn't, i.e., show that $X - Y = \phi$. This answers the first question in the list above.

The second and third questions are trivial to answer when given regular expressions. If the regular expression is not ϕ, then at least one string matches the expression. If the regular expression has a Kleene star, then it describes an infinite language.

One way to determine whether a *finite automaton* accepts any strings at all is to see if there is a path from the start state to an accepting state. This is often easy to do by inspection, but, as always, we seek an *algorithm* to answer the question; a machine with many states may not easily yield a solution by mere inspection.

To find such a path, we begin by "marking" the start state, and then following the procedure below:

1. For every marked state, s:
    a. Mark all states reachable from s
    b. Remove s and any edges connected to it
2. Repeat step 1 until an accepting state has been reached or until no marked states remain

The language is empty if and only if an accepting state is *not* reached.

Example 4–6

We illustrate the marking procedure with the automaton in Figure 2–9, repeated below with the initial state marked.

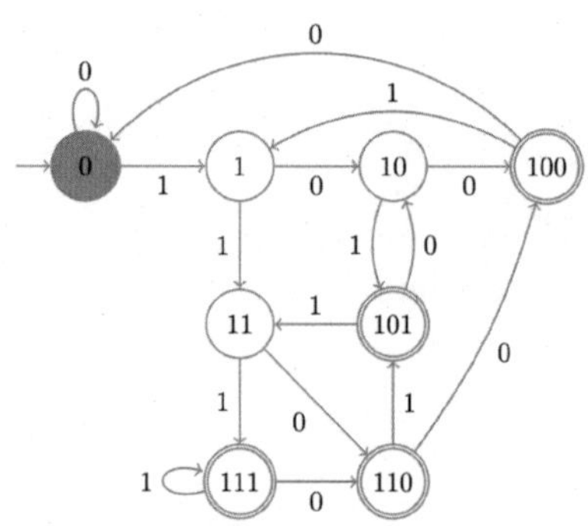

Figure 4–7: Searching Figure 2–9 for an accepting path

We now follow state 0's outgoing edges, marking the destination state (1), and delete any trace of state 0.

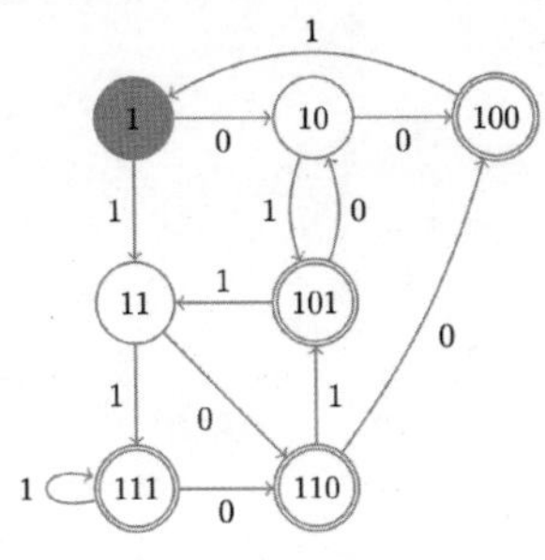

Figure 4–8: Step 2, after removing state 0

State 1 then leads us to states 10 and 11:

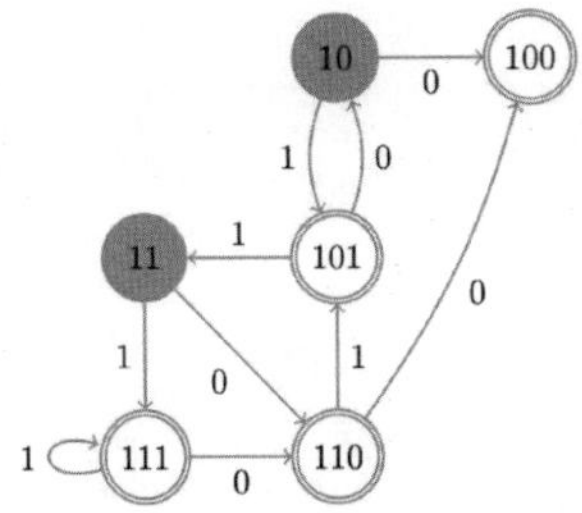

Figure 4–9: Step 2, after removing state 1

Finally, states 10 and 11 reach all the accepting states (although we only need to reach one to determine that the machine accepts some non-empty string). See Figure 4–10. We conclude that this automaton accepts a non-empty string.

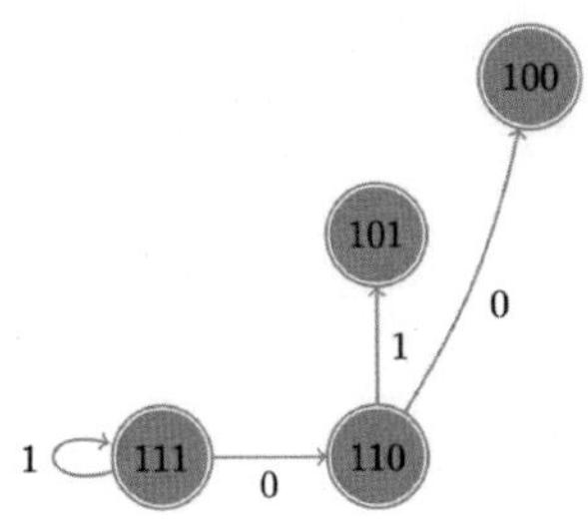

Figure 4–10: Final step reaching an accepting state

The procedure we have just described is a **depth-first traversal** of a directed graph, which visits all nodes exactly once. To implement a depth-first algorithm in Python, we define the following class:

A Node class modeling DFA states

```
class Node:
    def __init__(self, name, accept=False):
        self.name = name
        self.accept = accept
        self.children = []
        self.visited = False
```

Instances of the Node class track their name, acceptability, the other nodes they traverse to (their "children"), and whether they have been visited yet or not. We model the DFA in Figure 4–7 with the following Python code:

Modeling the DFA in Figure 2–9

```
# Create nodes
a = Node('0')
b = Node('1')
c = Node('10')
d = Node('11')
e = Node('100', True)
f = Node('101', True)
g = Node('110', True)
h = Node('111', True)

# Build graph
a.children = [a,b]
b.children = [c,d]
c.children = [e,f]
d.children = [g,h]
e.children = [a,b]
f.children = [c,d]
g.children = [e,f]
h.children = [h,g]
```

The following function, empty, implements a depth-first traversal and returns whether or not the language is empty (ignoring the empty string, of course).

Depth-first traversal searching for accepting states

```
def empty(marked):
    while marked:
        node = marked.pop()
        print('visited', node.name)
        node.visited = True
        if node.accept:
            return False      # Found an accepting state => not empty
        for c in node.children:
            if not c.visited:
                marked.append(c)
    return True
```

This function receives a list representing a stack of states that have been marked (called initially containing the start state). It marks each visited state, follows its edges, and adds to the stack for subsequent processing any unvisited states reached. It then erases any evidence of the current state being processed from further consideration. The procedure ends when it either encounters an accepting state or runs out of states to process (i.e., the stack is empty). The result of searching the graph follows.

Output for the search procedure

```
print ('Empty =', empty([a]))

''' Output:
visited 0
visited 1
visited 11
visited 111
Empty = False
'''
```

The language accepted by this automaton has been found *not* to be empty.

There are two more algorithms for determining if an automaton accepts any strings. One is to convert the automaton to a regular expression. If a non-empty regular expression results, it describes the pattern of the string(s) accepted, as explained earlier.

The procedure to test a language for infiniteness is based on the number of states in the automaton. We may assume, without loss of generality, that we are dealing with a minimal automaton with p

states. If a machine accepts any strings at all, it must accept a string of length p or less. To see why this is so, consider any accepting path through the machine. If the path has length p or greater, then that path has a cycle. This follows from the fact that the number of states in a path is one greater than the length of the string that path represents. A string of length p or greater has visited at least $p+1$ states, and since there are only p states to choose from, at least one state has been visited at least twice[3], thus forcing a cycle. Furthermore, any cycle in a DFA is of length no more than p, since there are only p states to choose from.

Consider the following DFA.

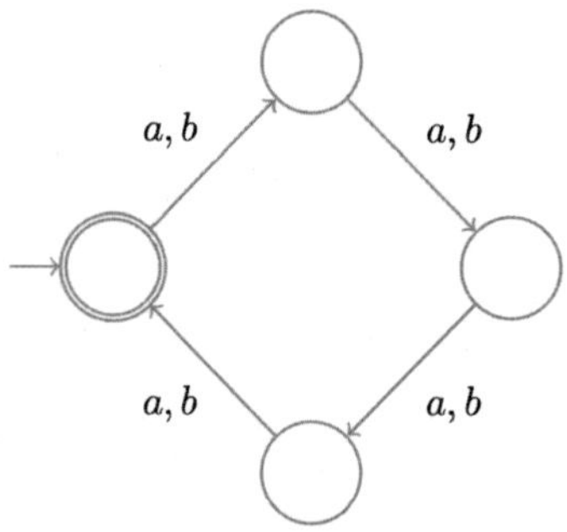

Figure 4–11: Accepts string of length $4n$

Since this machine is a single cycle in and of itself, and only the initial state accepts, the length of the smallest non-empty strings is equal to the number of states (namely, one of the sixteen strings matching the pattern $(a+b)^4$).

If a string in a regular language is at least p in length, then its accepting path has at least one cycle by the reasoning above. Since cycles are always between 1 and p in length, we can ignore all cycles and still find an accepting path of length no more than p.

Finally, if the DFA has no cycles, it must accept only strings of lengths in the range $[0, p-1]$, if it accepts any strings at all.

We now have a straightforward, if tedious, procedure for determining whether a DFA accepts any strings:

> To determine by computer whether a DFA accepts any strings, try all possible strings in Σ^* with length in the range $[0, p]$, where p is the number of states.

This would not be feasible to do by hand, but is easily done by computer.

[3]This follows from the *Pigeonhole Principle.*

Key Terms

depth-first traversal • pigeonhole principle

Exercises

1. *Prove* whether or not the following two DFAs accept the same language.

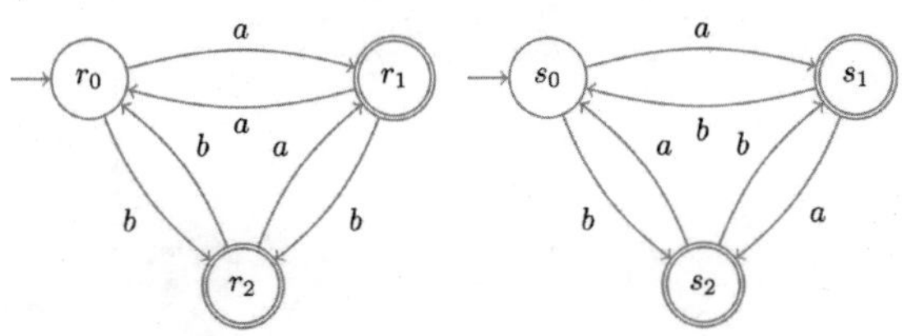

2. *Prove* whether or not the following two DFAs accept the same language.

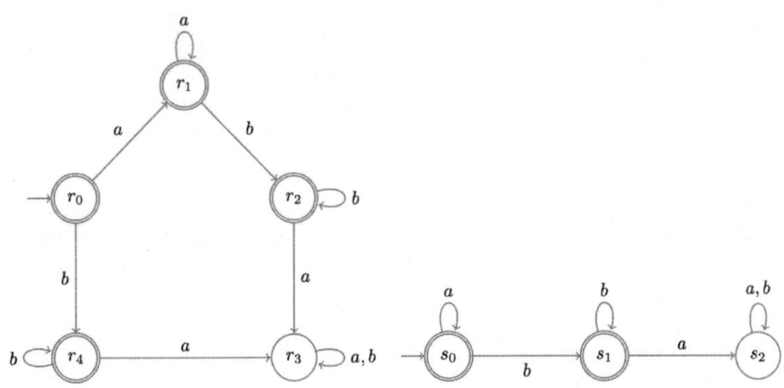

3. *Prove* whether or not the language accepted by the first DFA below is a subset of the language of the second DFA.

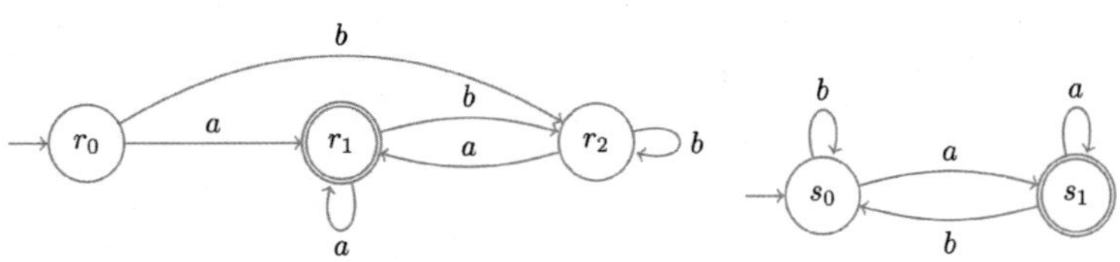

4. Does the following DFA accept any strings? *Prove* your result.

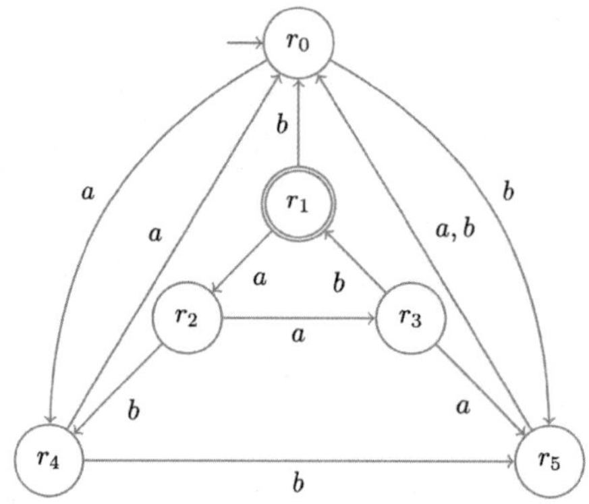

4.3 Infinite Regular Languages and a "Pumping Theorem"

The reasoning at the end of the previous section also leads to an algorithm for detecting whether a DFA's language is infinite. If a DFA accepts a string of length greater than or equal to p, where p is the number of states in the DFA, we know there is a cycle in an accepting path, so the language is infinite. We could therefore test all strings in Σ^* of length $p, p+1, p+2...$ for acceptance. The question is, when can we stop and give an answer? Since no cycle can contain more than p states, we only have to test strings of the following p lengths: $p, p+1, p+2, ..., 2p-2, 2p-1$. If a language accepts a string of length $2p$, we can ignore one of its cycles, which is no longer than length p, meaning that a string whose length is in the range $[p, 2p-1]$ must also be accepted.

> To determine by computer if the language of a finite automaton is infinite, it is sufficient to test all strings in Σ^+ with lengths in the range $[p, 2p-1]$ for acceptance.

As stated earlier you could also convert the language's automaton to a regular expression. If a Kleene star is present, then the language is infinite.

> To determine if the language of a finite automaton is infinite, convert the automaton to a regular expression and look for a Kleene star.

A "Pumping Theorem" for Infinite Regular Languages

Suppose a regular language, L, accepts a string, s, where $|s| \geq p$, and where p is the number of states in its associated minimal DFA. We know by the pigeonhole principle that by the time we have read the p-th symbol of s, a cycle has been found in an accepting path. Let's call the substring representing the *first* cycle found, y (traversing the cycle *once*), and the substring leading from the initial state to the first state of the cycle, x (which could be empty, as in Figure 4–14). Then we can represent s as the concatenation of three strings, xyz, where x and y are as just described, and z represents whatever is left in the string after y all the way to where the string ends in an accepting state. (See Figure 4–15 below.) The string z may contain other traversals of y's cycle and any other edges and cycles that may occur in the string after y. Like x, z may also be empty (as in Figure 4–14).

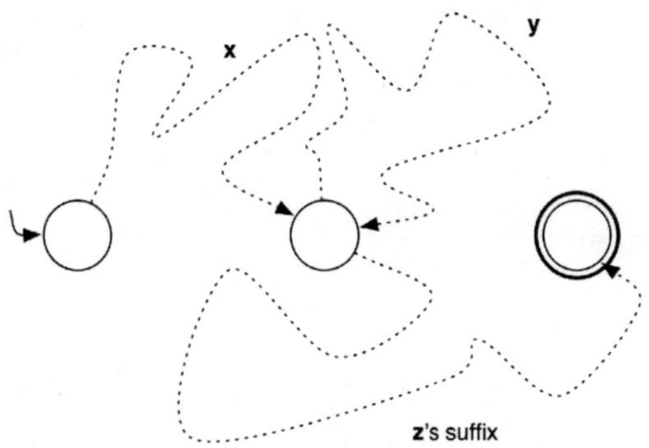

Figure 4–15: Partitioning s into xyz

The lines are dotted to suggest that many states may appear along the way. The first cycle (there may be others, which are part of z), represented by y, begins and ends with the state depicted in the middle above. The point here is that y represents the path of the *first cycle* traversed when processing an arbitrary string s, where $s \in L$ and $|s| \geq p$.

Example 4–7

Consider the regular language $ab(ba)^*b$, which the following incomplete DFA recognizes (the jail has been omitted for clarity).

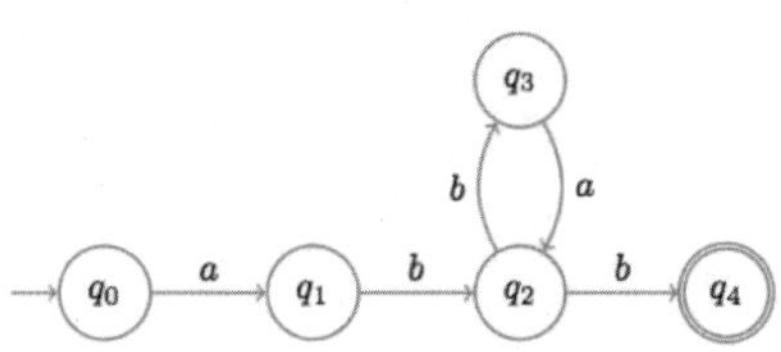

Figure 4–16: Accepts $ab(ba)^*b$

Here we have $x = ab$, $y = ba$, and z always ends with b. z will vary depending on which string in the language we are considering. The following table shows the partitions for strings in this language.

Table 4–7: Partitioning $ab(ba)^*b$ into xyz

String	Partitioning	z
abb	xb	b
$ab\underline{ba}b$	xyz	b
$ab\underline{baba}b$	xy^2b	$yb = bab$
$ab\underline{bababa}b$	xy^3b	$y^2b = babab$
...	...	...
$ab\underline{(ba)^n}b$	xy^nb	$y^{n-1}b = (ba)^{n-1}b$

The string abb does not use the $q_2 \to q_3 \to q_2$ cycle, but the other strings *must*, since they are of length $p = 5$ or more (any string at least p in length forces a cycle—in other words, a cycle must occur within the first p input symbols, expressed mathematically as $|xy| \leq p$). Finally, we can repeat, or "pump", the cycle, y, at will, and still end up with a string in the language by appending z.

We can express these ideas as a theorem[4]:

The Pumping Theorem for Regular Languages

For any infinite regular language, L, there is a positive number, p, such that for every string, s, where $s \in L$ and $|s| \geq p$, we can partition the string as $s = xyz$, and the following hold:

1. $|y| > 0$
2. $|xy| \leq p$
3. $xy^*z \in L$

The number, p, is the number of states in the minimal DFA for the language. The substring y is non-empty because any cycle has at least one edge. The Kleene star on y in condition 3 above represents the fact that we can choose to follow the cycle defining y zero or more times (we can ignore the cycle) and still reach an accepting state via z, thus obtaining other valid strings in the language.

Why do we care about this "pumpability" of regular languages? Because if all infinite, regular languages must be "pumpable", and we can show that if a given, infinite language is *not* pumpable (i.e., it does *not* satisfy the Pumping Theorem), then we can conclude that that language is *not* regular. This leads us to another category of languages to be introduced in the next chapter.

It is important to remember what this theorem does *not* say. It does not say that if a language is pumpable, then it is regular. That is the *converse* of the theorem, which is not logically equivalent to

[4]The term "Pumping Lemma" is also commonly used instead of Pumping Theorem.

the original proposition. But the *contrapositive*, if a language is not pumpable then it is not regular, *is* logically equivalent to the original proposition of the theorem (i.e., $p \to q \equiv \neg q \to \neg p$). This is how we use the Pumping Theorem.

We use the Pumping Theorem to show that a formal language is *not* regular. We *cannot* use the Pumping Theorem to show that a language is regular.

Example 4–8

We now show that the language $L_{ab} = a^n b^n$ is not regular by showing that it does not satisfy the Pumping Theorem.

Intuitively we can surmise that L_{ab} is not regular because a finite machine has no way to track an arbitrarily large number of a's to match the subsequent b's; we can only go so far before we run out of states (review Exercise 3.1-7). Consider the diagram below accepting the singleton language $\{ab\}$. (Once again we omit the jail state for readability.)

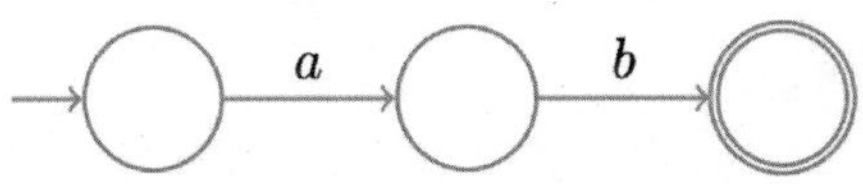

Figure 4–17: Accepts $\{ab\}$

Now, how can we define a machine to accept $\{ab, aabb\}$? Or $\{ab, aabb, aaabbb\}$? We need to add more states. (See the next two diagrams.)

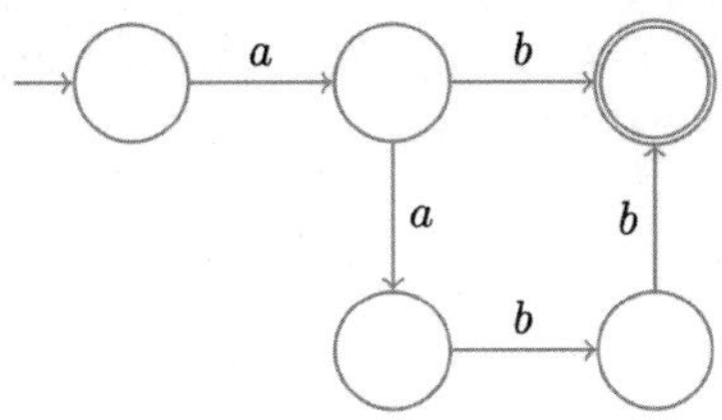

Figure 4–18: Accepts $\{ab, aabb\}$

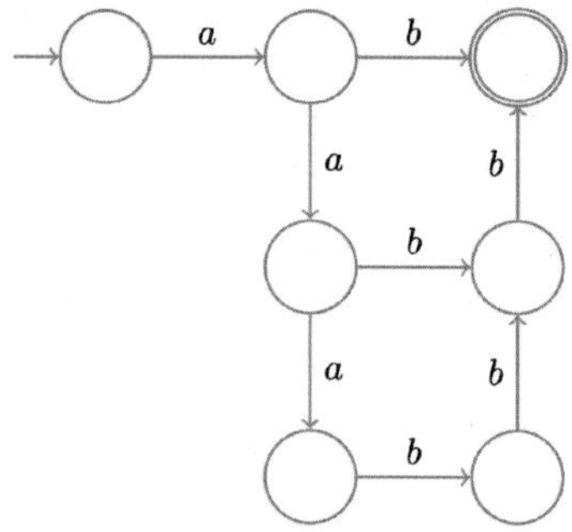

Figure 4–19: Accepts $\{ab, aabb, aaabbb\}$

A machine with a *finite* number of states will never have enough states to match an arbitrarily large number of symbols.

The reasoning above satisfies our intuition, but it is *not a proof*! If $L_{ab} = a^n b^n$ were regular, then per the Pumping Theorem, we could pick *any string* in this language of at least p symbols in length, and partition it into three concatenated substrings, xyz, and then pump y to our heart's content and still obtain a string in the language, as described above. This is how infinite regular languages behave. We will choose the string $a^p b^p$, which is more than long enough for us to invoke the Pumping theorem, and show that it cannot be so partitioned.

If L_{ab} were regular, then, by the Pumping Theorem, we could partition $a^p b^p$ into three parts, xyz. Since the condition $|xy| \leq p$ must hold[5], we know that the substring y must occur somewhere within the first p symbols of $a^p b^p$, which means that it occurs somewhere within the initial run of a's in this string (it doesn't matter exactly where or how long it is; all we know is that it is no longer than p). It suffices to say that in our case, y consists of nothing but a's, because there are p a's to begin with. Now, what happens when we try to pump y? The number of a's in the string changes, but the number of b's does not. This causes the pumped string to "fall out" of the language—it no longer has an equal number of a's and b's. This violates the Pumping Theorem (this language doesn't "pump"), so we conclude that L_{ab} is not regular. (That was easy!)

The key to handily solving this problem was how we chose the test string. The Pumping Theorem applies to *any string* at least p symbols long, because such a string forces a cycle when processing the first p symbols in an accepting path in the automaton. We chose a string that allowed us to use the condition $|xy| \leq p$ to our advantage. It wasn't necessary to do this, however, for this language. Suppose we chose a string $a^k b^k$ where $2k \geq p$, but k is not necessarily greater than or equal to p. In this case the condition $|xy| \leq p$ would serve no useful purpose (because y could have a's and/or b's), but we could still consider all the ways in which y might appear:

[5]Remember: a cycle must occur within the processing of the first p input symbols of any string in a regular language.

1. If y appears wholly within the initial run of a's, then we conclude, as before, that the number of a's changes, but the number of b's does not, violating the Pumping Theorem (the number of a's no longer equals the number of b's).
2. If instead, y straddles the boundary between the a's and b's, then it is of the form $a^i b^j$, so we have $a^k b^k = a^{k-i} a^i b^j b^{k-j}$. Pumping y yields a string that is no longer a single run of a's followed by a single run of b's. For example, the string xy^2z is $a^{k-i}(a^i b^j)(a^i b^j)b^{k-j}$ (parentheses shown to emphasize y^2), which has *four runs* of characters: $a^k b^j a^i b^k \notin L_{ab}$. Again, pumping fails.
3. The only other possibility is that y occurs wholly inside the run of b's, which is analogous to case 1 above. Pumping in this case changes the number of b's without changing the number of a's, again falling out of the language L_{ab}.

So, without using the condition $|xy| \leq p$, we have to consider all possible cases of how y can appear in the string to show that there is **no partition whatsoever** that satisfies the Pumping Theorem. The second two steps above were unnecessary when we used the string $a^p b^p$ along with Condition #2 of the Pumping Theorem. It is a good idea to use condition 2 of the Pumping Theorem whenever it applies (which is often).

We *cannot assign a fixed, numerical length* to x, y, or z, nor say precisely which positions the partitions occupy in the string. All we may assume are the conditions of the Pumping Theorem, which are in terms of the parameter p, and we must show that **no valid choice** for y satisfies the Pumping Theorem. It is *always erroneous* to say, for example, "assume that $|y| = 2$", etc. (the same goes for x and z).

We may *only assume* that $|xy| \leq p$ (and therefore, $|y| \leq p$), and that y must appear within the leading p symbols of a string to be pumped.

Example 4–9

A palindrome is a string that reads the same backwards and forwards, like "radar" and "madam." The language of palindromes over $\Sigma = \{a, b\}$ is not regular. Once again, our intuition tells us that we would need an infinite number of states to check for matching symbols on either end of an arbitrarily large string. To prove that the language is not regular, we need to show that it doesn't "pump". To make the process as easy as possible, we pick a palindrome that has the same symbol occupying the first p symbols in the string, such as $a^p b a^p$, and equate it to the expression xyz. The requirement $|xy| \leq p$ implies that the substring y must appear within the first p symbols, which are all a's. But this means that when we pump y i-times, say, we get a string of the form $a^{p+i|y|} b a^p$, where the number of a's before the b has changed, but the number of a's after the b hasn't, resulting in a *non-palindrome*. This violates the requirements of the Pumping Theorem for regular languages, so we conclude that this language is not regular.

Example 4–10

The language of repeated halves, $\{ww \mid w \in (a+b)^*\}$, is not regular. An example of a string in this language is a^pba^pb. The same logic from the previous example applies here. The substring y must occur within the initial run of a's. When it is pumped, the number of initial a's changes, but nothing else changes, giving a string of the form $a^{p+i|y|}ba^pb$, which is *not* the concatenation of two equal substrings. Pumping fail.

Example 4–11

The language over $\Sigma = \{(,)\}$ of *balanced parentheses*, with strings such as $(()(()))$, is not regular. If it were, then we could pump the string $(^p)^p$. But, once again, the fact that $|xy| \leq p$ means that y consists of only left parentheses, which, when pumped, destroys the balance.

With simple examples like the foregoing one might get the idea that it is always easy to show that a language is non-regular. We now examine three more languages that present increasing levels of difficulty in proving them non-regular.

Example 4–12

The language $\{a^mb^n \mid m > n\}$, meaning that there are always more a's than b's, is not regular. If it were, then we could pump the string $a^{p+1}b^p$. This time, we take advantage of the Kleene star in the condition $xy^*z \in L$ from the Pumping Theorem by using the *zero-repetitions* case. We know that y must occur within the run of a's. When we *pump down* once (i.e., ignoring the cycle that defines y), we obtain the string $a^{p-|y|+1}b^p$, making the number of a's less than or equal to the number of b's—another pumping fail. Pumping "up" in this case would not accomplish anything.

For some languages you need to *pump down* once to show a language doesn't satisfy the Pumping Theorem.

Example 4–13

The language PRIME = $\{a^n\}$ where n is a *prime number* is not regular. The alphabet here is the singleton $\Sigma = \{a\}$. Let us consider the string a^k where $k \geq p$ and k is prime. Our goal is to pump y up a certain number of times that will guarantee that the resulting string is a *composite number* instead of a prime. The only fixed quantity we have is k, so we'll pump y k-times getting the string $xy^{k+1}z$. Since the string consists only of a's, we can rearrange the previous expression as $xyzy^k$ by moving all but one of the y's to the right. Recalling that $|xyz| = k$, we have $|xyzy^k| = |xyz| + |y^k| = k + k|y| = k(1 + |y|)$. The length of our pumped string is no longer a prime number, failing the Pumping Theorem.

Example 4–14

The language $\{a^{n^2} \mid n > 0\}$ is not regular. The alphabet here is again $\Sigma = \{a\}$. We know that the distance between two adjacent perfect squares increases by 2, because finding the next square is obtained by adding the next larger odd number (the first square is 1, the next is $1 + 3 = 4$, then $4 + 5 = 9$, etc.). In general we have

$$(n + 1)^2 = n^2 + (2n + 1)$$

where the parenthesized expression on the right is the next odd number. In other words, the difference between nth and $(n + 1)$th perfect square numbers is $2n + 1$, which is *not a fixed quantity.* It is therefore impossible to always obtain a perfect square by repeatedly adding a fixed number such as $|y|$.

To mathematically show that this language fails the Pumping Theorem, we consider the string a^{p^2}, and use the condition $|y| \leq p$. Pumping up once from a^{p^2} gives the string $a^{p^2+|y|}$. We now have

$$|a^{p^2+|y|}| = p^2 + |y| \leq p^2 + p < p^2 + 2p + 1 = (p + 1)^2$$

The fact that $|y| \leq p$ and that $p < 2p + 1$ shows that adding $|y|$ a's didn't quite make it up to the next perfect square.

We can also show that this language is not regular by pumping down and failing to reach $a^{(p-1)^2}$ if we make the reasonable assumption that $p > 1$:

$$p^2 - |y| \geq p^2 - p = p^2 - 2p + p > p^2 - 2p + 1 = (p - 1)^2$$

Sometimes the condition $|y| \leq p$ is key to show that a language is not regular using the Pumping Theorem.

Using Closure Properties to Show Non-Regularity

Now that we know a few non-regular languages, we can conclude that other languages are non-regular merely by using the *closure properties* of regular languages. The language NOTPRIME ($\{a^n\}$ where n is *not* prime), for example, is not regular. If it were, then its complement, PRIME, would be also, but we showed in Example 4–13 that PRIME is *not* regular, so NOTPRIME can't be regular, since regular languages are closed under complement. QED.

It is fruitless to attempt to use the Pumping Theorem on a language like this, defined with a negative (e.g., using "not"). Instead, we show its *complement* is not regular. Depending on the language, other closure properties can be used to show that a language is not regular.

We can also show that the language EQUAL, defined as $\{n_a(w) = n_b(w) \mid w \in (a+b)^*\}$ (i.e., strings where the number of a's and b's are equal, but the letters can appear in any order), is not regular by applying the Pumping Theorem to the string a^pb^p (remember, we can choose *any* qualifying string in the language). But we can use closure properties of regular languages to avoid having to use the Pumping Theorem. In this case, consider the set of strings formed by the *intersection*, EQUAL $\cap\ a^*b^*$. The is the set of strings of a's and b's where the number of each letter is equal, but also where the a's must precede the b's. But this is precisely the language $L_{ab} = a^nb^n$, which we already know is *not* regular. Since a^*b^* represents a regular language, then EQUAL must not be regular, because regular languages are closed under intersection.

Example 4–15

The language $L_{ne} = \{a^mb^n \mid m \neq n\}$ is not regular (notice the $\neq$, which implies a "not"). Since we don't get to choose the size of y, it is possible that pumping y will leave the number of a's unequal to the number of b's. In general, we can't prove a negative with any theorem directly. This language is *not the complement* of $L_{ab} = \{a^nb^n\}$, by the way, because $\overline{L_{ab}}$ also includes strings that aren't necessarily a run of a's followed by a run of b's, no matter the number of occurrences of each character (strings like $bbaa$, as well as strings like baa, and $abab$). L_{ne} *does* include all the strings in $\overline{L_{ab}}$, however, so we can compute the following difference:

$$a^*b^* - L_{ne} = \{a^nb^n\} = L_{ab}$$

We have removed the elements of L_{ne} from the regular language a^*b^* and obtained a language that is not regular, so L_{ne} can't be regular either, since regular languages are closed under difference.

Key Terms

contrapositive • non-regular language • Pumping Theorem for regular languages

Exercises

1. Using the Pumping Theorem, show that the language $a^m b^m$ where m is a perfect cube, is not regular.
2. Show that the language $\{n_a(w) \neq n_b(w) \mid w \in (a+b)^*\}$ is not regular.
3. Show that the language $\{a^m b^n\}$ where $\frac{m}{n}$ is a positive integer is not regular.
4. Show that the language $\{a^{2^n} \mid n > 0\}$ is not regular.

Chapter Summary

We have concluded our study of the class of regular languages, which are closed under more operations and amenable to more decision algorithms than the other classes of languages we will investigate. Knowing that a language is regular allows us to utilize a *finite automaton* that recognizes it, which in turn makes writing applications that use the language straightforward. Likewise, knowing that a language is not regular saves us the trouble of trying to find a finite automaton or regular expression that describes it, since we know that there isn't one.

Any language that needs to track arbitrarily large counts of characters cannot be regular, since we have only a finite number of states with which to track them. To *prove* that a language is not regular, however, we must use the **Pumping Theorem** (which takes a lot of practice to become proficient at) or the closure properties of regular languages in conjunction with previously established, non-regular languages.

The Pumping Theorem is easier to grok if you keep in mind its simple logic:

- long strings in a regular language force a *cycle* in its automaton
- a cycle occurs within the first p characters of any such string
- the cycle is no longer than p characters in length
- that cycle can be repeated or ignored altogether
- any string so formed will be in the language, *if the language is regular.*

The most crucial part of using the Pumping Theorem is usually finding a convenient test string. And never forget: you *don't* use the Pumping Theorem to show that a language is regular; **pumpability does not imply regular**! To show that a language is regular requires finding a finite automaton, regular expression, or regular grammar that represents it.

II Context-Free Languages

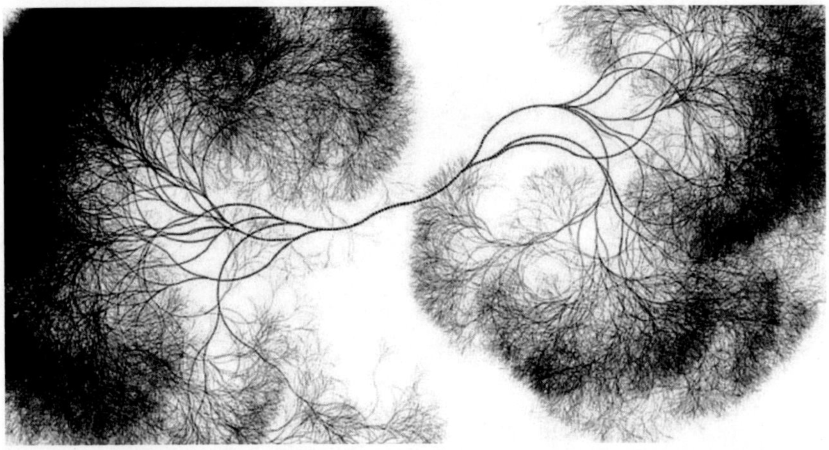

Context-Free Art by Anthony Mattox, spacecollective.org, used with permission

5. Pushdown Automata

We must overcome the notion that we must be regular... it robs you of the chance to be extraordinary and leads you to the mediocre.

– Uta Hagen

Rock Stack, by Deidre Woollard Licensed under CC-BY 2.0

Where Are We?

Language Class	Recognized By	Generated By
Regular	Finite Automata	Regular Grammar Regular Expression
Context-free	Pushdown Automata	Context-free Grammar
Recursively Enumerable	Turing Machine	Unrestricted Grammar

Chapter Objectives

- *Understand the effect of adding a stack to finite automata*
- *Design pushdown automata for context-free languages*
- *Distinguish between non-deterministic and deterministic pushdown automata*

Having discovered that there are some languages that are not regular, we now look at a type of machine that accepts some of these languages. The type of machine we will use is essentially identical to the finite automata we have seen, except we add the use of an external *stack* data structure as auxiliary storage, and is aptly named a **pushdown automaton** (PDA). The class of languages pushdown automata accept are the *context-free* languages (CFLs). We will also see in this chapter that, unlike regular languages, there is a difference between non-deterministic and deterministic CFLs. Some CFLs are inherently non-deterministic.

5.1 Adding a Stack to Finite Automata

With a stack at our disposal, we have the option of popping and/or pushing symbols from/onto the stack as we process an input stream. Transitions are taken depending on the current state, the symbol read, *and* the symbol on the top of the stack. When following a transition, we can not only change state, but we can pop one symbol and push one or more symbols onto the stack. Transition edges in PDAs take the form `<read-char>,<pop-char>` $\rightarrow$ `<push-string>`. The following figure shows a PDA that accepts the language $\{a^n b^n \mid n > 0\}$.

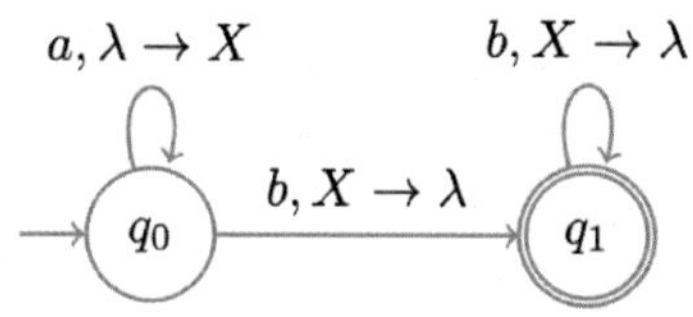

Figure 5–1: PDA accepting $\{a^n b^n \mid n > 0\}$

We always start with an *empty stack*. The idea with the PDA above is to add an X to the top of the stack as a *counter* every time we read an a. Whenever we read a b, the applicable transition indicates that we must pop an X from the stack. The PDA accepts only if it ends in an accepting state *and* the stack is empty.

The transition rule $a, \lambda \rightarrow X$ means[1], "if an a is read from the input stream, don't pop anything from the stack and push an X onto the stack." The presence of a lambda means to ignore reading, pushing, or popping, depending on where it appears in the transition rule. The symbol before the arrow represents the *symbol* to pop, and the string after the arrow represents the *string* of symbols to push. If the expected symbol is not present to be popped, then that transition rule cannot be applied. If there is no applicable rule, the machine "crashes" (similar to going to a jail state) and rejects the input. We rarely include jail states in PDAs.

> Processing a string in a PDA must end in an *accepting state* with an *empty stack* to accept an input string. At most *one symbol* at a time may be *popped* from the stack.

The following table traces the processing of the input string $aabb$ through the PDA in Figure 5–1.

[1]We use the right-arrow because in general, we are replacing a symbol on the top of the stack with whatever symbols we are pushing, like the replacement rules in grammars, which also use right-arrows. If you find arrows tedious to draw by hand, you can use a slash, as in $b, X/\lambda$, or just a comma: b, X, λ.

Table 5–1: Tracing the string $aabb$

State	Input	Stack
q_0	$aabb$	λ
q_0	abb	X
q_0	bb	XX
q_1	b	X
q_1	λ	λ

The top of the stack is the leftmost character in the string representing the stack in the third column. The goal is to consume the string and end in q_1 with an empty stack, the latter constituting a rudimentary counting mechanism. The following Python code implements this PDA.

Python implementation of the PDA in Figure 5–1

```
def anbn(s):
    state = 'q0'
    stack = list()
    for c in s:
        if state == 'q0':
            if c == 'a':
                stack += 'X'
            elif c == 'b':
                try:
                    top = stack.pop()
                except:
                    return False
                assert top == 'X'
                state = 'q1'
            else:
                return False
        elif state == 'q1':
            if c == 'b':
                try:
                    top = stack.pop()
                except:
                    return False
                assert top == 'X'
            else:
                return False
    return state == 'q1' and not stack

s = 'aabb'
print(s,'accepted' if anbn(s) else 'not accepted')      # aabb accepted
```

We can also use a symbolic notation instead of a trace table to represent the sequence of steps in processing an input string with a PDA. The turnstile symbol, $\vdash$, sometimes pronounced "yields," indicates the progress of a single step in the machine. The symbolic notation for the trace in Table 5-1 appears below.

$$(q_0, aabb, \lambda) \vdash (q_0, abb, X) \vdash (q_0, bb, XX) \vdash (q_1, b, X) \vdash (q_1, \lambda, \lambda)$$

As a shorthand, we express the entire movement through the machine applying the Kleene star to the turnstile symbol as $(q_0, aabb, \lambda) \vdash^* (q_1, \lambda, \lambda)$. The pair of symbols $\vdash^*$ means "yields in multiple steps." The goal of a PDA is to start in the configuration (q_0, s, λ) for input string, s, and end in a final state, q_f, say, with the configuration, (q_f, λ, λ)—in other words, $(q_0, s, \lambda) \vdash^* (q_f, \lambda, \lambda)$.

We now present the formal definition of a pushdown automaton.

Definition 5.1

A **pushdown automaton** (PDA) is a finite state machine, with an associated stack, consisting of the following:

- Q, a set of states, one of which is the initial (or start) state, and a subset of which are final (or accepting) states.
- Σ, an input alphabet of symbols.
- Γ, a **stack alphabet** (which can include Σ).
- $\delta : Q \times (\Sigma \cup \{\lambda\}) \times (\Gamma \cup \{\lambda\}) \rightarrow$ *finite subsets of* $Q \times \Gamma^*$, a **relation** which, given a state, an optional input character from Σ, and an optional stack character from Γ, determines the next state of the machine and the string from Γ^* to push onto the stack.

The input and stack alphabets do *not* have to be mutually exclusive. Also, δ is a *relation*, allowing multiple outputs in $Q \times \Gamma^*$ for the same input, so this definition accommodates *non-determinism*. We can formally define the transition relation of the PDA in Figure 5–1 as follows:

$$\begin{aligned}\delta(q_0, a, \lambda) &= (q_0, X)\\ \delta(q_0, b, X) &= (q_1, \lambda)\\ \delta(q_1, b, X) &= (q_1, \lambda)\end{aligned}$$

or in set notation:

$$\delta = \{((q_0, a, \lambda), (q_0, X)), ((q_0, b, X), (q_1, \lambda)), ((q_1, b, X), (q_1, \lambda))\}$$

This PDA happens to be deterministic, so δ in this case is a *function.*

Example 5–1

The following PDA accepts $\{a^n b^{2n} \mid n \geq 0\}$.

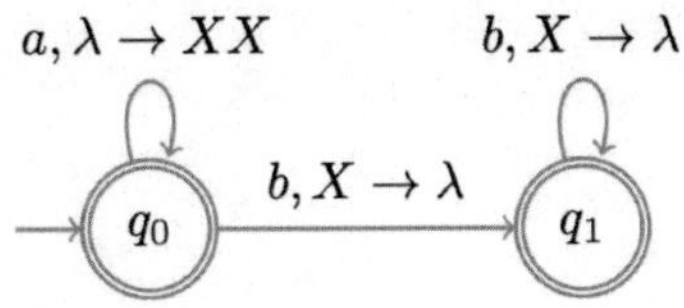

Figure 5–2: PDA accepting $\{a^n b^{2n} \mid n \geq 0\}$

For every a we push two X's onto the stack, so it takes twice as many b's to empty the stack. Both states are accepting, since n can be zero, but remember that the *stack must be empty* for a string to be accepted.

Example 5–2

The following PDA accepts the language $\{wcw^R \mid w \in (a+b)^*\}$, that is, odd-length palindromes of a's and b's with a c in the middle.

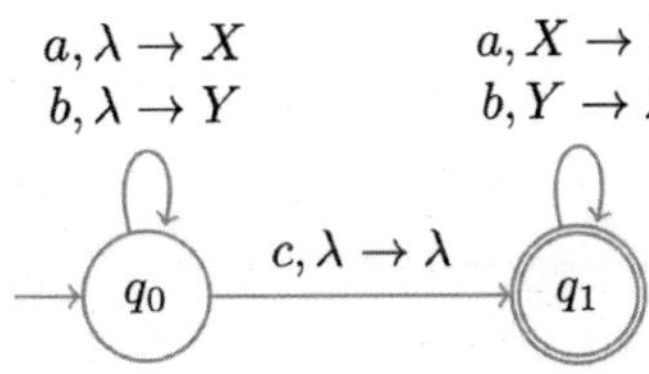

Figure 5–3: Accepts $\{wcw^R \mid w \in (a+b)^*\}$

We push distinct symbols for a's and b's. Once the c is consumed, we expect to find the *reversal* of the first half of the string in terms of X's and Y's. The minimal string, c, is accepted, since it does not modify the stack (leaving it empty), and ends in an accepting state.

All the examples so far have been *deterministic* PDAs—there is no ambiguity as to which transition applies for each input symbol. It is not possible, however, to construct a deterministic PDA (a DPDA) for the language of *even-length* palindromes, since there is no symbol present to mark the middle of the string, and therefore no deterministic way to instruct the machine to move from state q_0 to state q_1. See Figure 5–4 below.

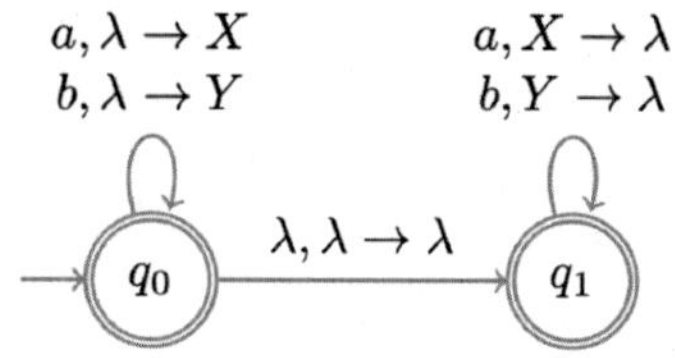

Figure 5–4: Accepts $\{ww^R \mid w \in (a+b)^*\}$

The trace for the string $abba$ would be:

$$(q_0, abba, \lambda) \vdash (q_0, bba, X) \vdash (q_0, baYX) \vdash (q_1, baYX) \vdash (q_1, a, X) \vdash (q_1, \lambda, \lambda)$$

The move from the q_0 to the q_1 must be made *non-deterministically.* We will have more to say about PDAs and determinism in the next section.

The following non-deterministic PDA (NPDA) accepts *all* palindromes over $\Sigma = \{a, b\}$:

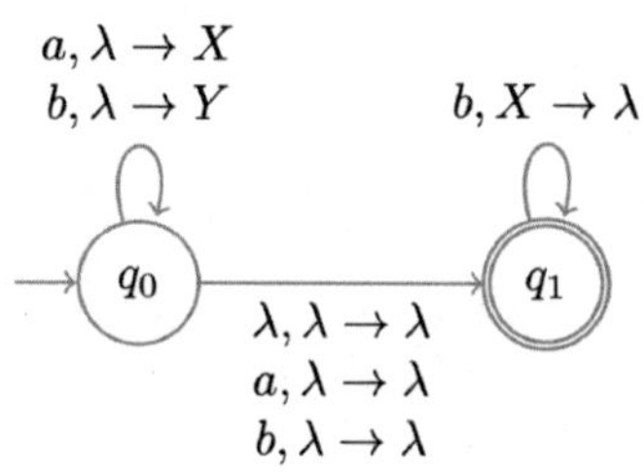

Figure 5–5: Accepts all palindromes

As before, the middle symbol, if present, must not affect the stack.

Example 5–3

The following NPDA accepts strings where the number of b's is between one and two times the number of a's by non-deterministically pushing one or two stack symbols.

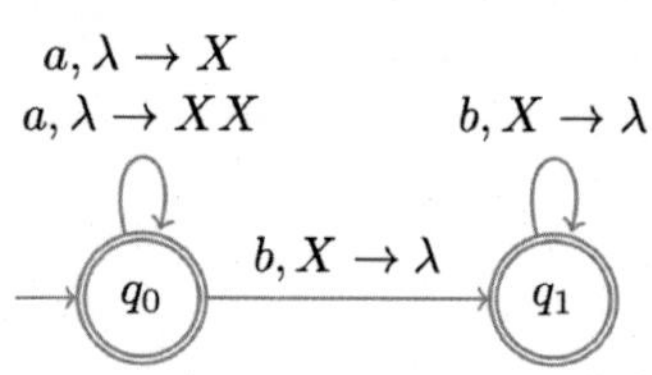

Figure 5–6: Accepts $\{a^m b^n \mid m \leq n \leq 2m\}$

One possible trace for the string $aabbb$ is:

$$(q_0, aabbb, \lambda) \vdash (q_0, abbb, X) \vdash (q_0, bbb, XXX) \vdash (q_1, bb, XX) \vdash (q_1, b, X) \vdash (q_1, \lambda, \lambda)$$

We also could have pushed two X's for the first a instead:

$$(q_0, aabbb, \lambda) \vdash (q_0, abbb, XX) \vdash (q_0, bbb, XXX) \vdash (q_1, bb, XX) \vdash (q_1, b, X) \vdash (q_1, \lambda, \lambda)$$

Example 5–4

The following PDA accepts the language L_{eq}, which contains strings having an equal number of a's and b's (again expressed by the notation $\{n_a(w) = n_b(w) \mid w \in (a+b)^*\}$).

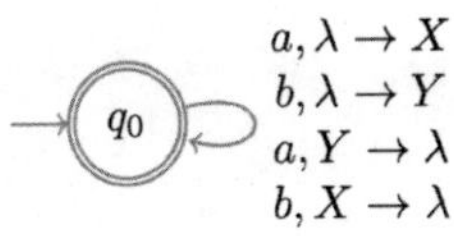

Figure 5–7: Accepts $\{n_a(w) = n_b(w) \mid w \in (a+b)^*\}$

Since the symbols can appear in any order, we use separate stack symbols for a and b, as we did in Example 5–2. When we read an a, and if there is a Y on top of the stack, we pop it; otherwise we push an X (and analogously for b). Here is a trace of the string *baaabb* (since there is only one state, we omit it in the table below, and instead indicate which transitions apply):

Table 5–2: Tracing the string *baaabb*

Input	Stack	Transition to Apply Next
baaabb	λ	$b, \lambda \to Y$
aaabb	Y	$a, Y \to \lambda$
aabb	λ	$a, \lambda \to X$
abb	X	$a, \lambda \to X$
bb	XX	$b, X \to \lambda$
b	X	$b, X \to \lambda$
λ	λ	(accept)

The transition rules must be used so that X's and Y's never appear together on the stack simultaneously (hence this is a non-deterministic machine). The rules in this PDA are designed so that the presence of X's on the stack means that more a's than b's have been read so far, and similarly for Y's and b's. Since there is a path through the machine that ends in an accepting state with an empty stack, the string is accepted.

Example 5–5

Try to guess what language the following PDA accepts before reading the next paragraph. (*Hint*: the transitions on the second state are the same as in the language L_{eq} above.)

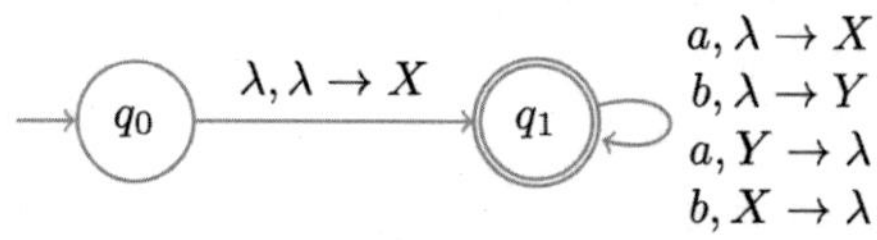

Figure 5–8: What language does this PDA accept?

We "seed" the stack with an X. Since X's are only removed by reading the symbol b, there must be one additional b than would be present if the initial X had not been pushed. This PDA, therefore, recognizes the language of all strings w over $\Sigma = \{a, b\}$ where $n_b(w) = n_a(w) + 1$. If the string begins with a b, then the initial X is popped, otherwise the initial X is popped when reading a b later on.

Example 5–6

To recognize the language $n_b(w) = 2n_a(w)$, we must consider how two b's can be associated with each a in a string. Either both b's precede the a's, or one b precedes and the other follows, or both b's follow their companion a. If there are two Y's at the top of the stack when we read an a, we can just pop them both to keep the count balanced. If the stack consists of only one Y, we can remove it, and then push an X for a subsequent b to remove. If the stack is empty or has an X at the top, then we add two X's for subsequent b's to remove. See the diagram below.

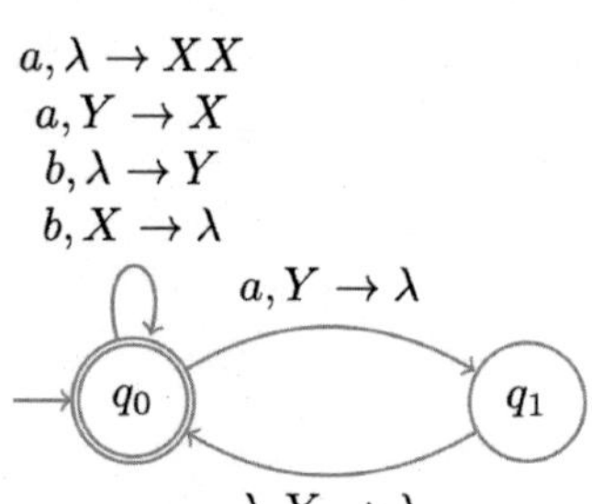

Figure 5–9: Accepts $n_b(w) = 2n_a(w)$

The following table illustrates all three cases by tracing the string *bbababab b* through the PDA.

Table 5–3: Tracing *bbababab*b

State	Input	Stack
q_0	*bbabababb*	λ
q_0	*bababb*b	Y
q_0	*ababab*b	YY
q_1	*bababb*	Y
q_0	*bababb*	λ
q_0	*ababb*	Y
q_0	*babb*	X
q_0	*abb*	λ
q_0	*bb*	XX
q_0	*b*	X
q_0	λ	λ

We have yet to show how to do a **transition tables** for PDAs. Though rarely used, they come in handy in one circumstance that appears in Chapter 7. A transition table for the PDA in Figure 5–9 appears below. *Note*: the columns are the possible `<read-char>,<pop-char>` input pairs that appear in the PDA. The result in each internal cell is the resulting state along with what will be pushed on the stack, where multiple results may exist as a set in curly braces. Many options are not possible, indicated by a dash.

Table 5–4: Transition table for the PDA in Figure 5–9

	a, λ	a, Y	b, Y	b, X	λ, Y
$+q_0$	q_0, XX	$\{q_0/X, q_1/\lambda\}$	q_0/Y	q_0/λ	–
q_1	–	–	–	–	q_0, λ

Remember that the *input* for the transition function in a PDA consists of three things: the current state, the input symbol (if used), and the symbol to pop from the top of the stack (if used). The first table entry above indicates that when in state q_0, if the input symbol is a, we can push the string XX onto the stack. The next entry to the right reflects the two choices available in q_0 when the input is a and there is a Y at the top of the stack. The missing entries in the table indicate `(<state>,<read-char>,<pop-char>)` combinations that do not appear in the machine—they are not valid transitions for this PDA (i.e., they implicitly lead to a jail state).

Context-free languages are closed under union, concatenation, and Kleene star using the same machine-combining strategies that we saw for regular languages. The following PDA accepts strings of the form $a^i b^j c^k$ where either the number of b's or c's (but not both simultaneously) are equal to the number of preceding a's, by allowing for the union of the respective corresponding machines.

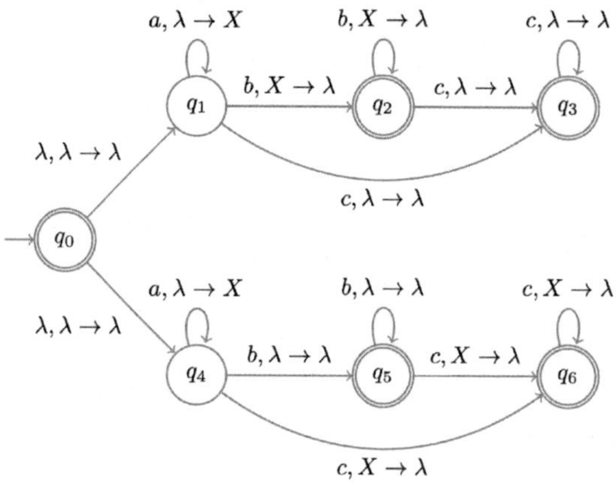

Figure 5–10: Accepts $\{a^i b^j c^k \mid i = j \vee i = k\}$

The top branch of the machine ignores the stack when processing c's, and the bottom machine ignores the stack when processing b's. Strings matching the patterns b^* and c^* are accepted, as they should be.

Example 5–7

Consider how to design a PDA for the language $L_{ne} = \{a^i b^j \mid i \neq j\}$. First, this is *not the complement* of the language $\{a^n b^n\}$! The complement also contains strings *not* of the form "a run of a's followed by a run of b's", such as *babbaab*. Furthermore, the class of non-deterministic context-free languages is *not* closed under complementation (as we will show later). However, we can see that L_{ne} can be also expressed as $\{a^i b^j \mid i < j\} \cup \{a^i b^j \mid i > j\}$. If we can find PDAs for these two sub-languages, we can take their union. For the first PDA, we will force at least one additional trailing b and for the second PDA, at least one additional a at the beginning. Neither of these extra inputs will use the stack. See the following two figures.

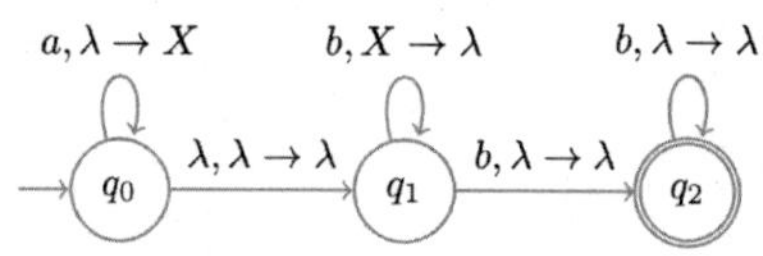

Figure 5–11: Accepts $\{a^i b^j \mid i < j\}$

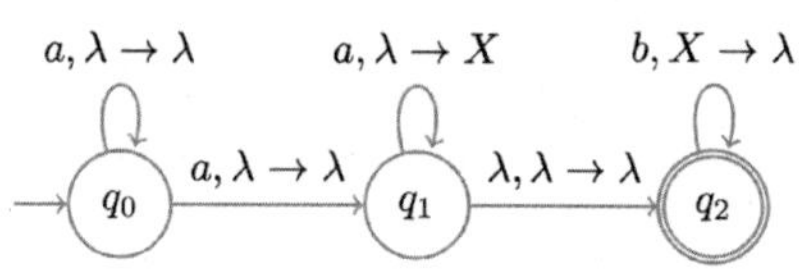

Figure 5–12: Accepts $\{a^i b^j \mid i > j\}$

The machine in the first of the two figures above accepts the string b, and the second machine accepts a. We can now form the union of these two PDAs by creating a new start state and moving non-deterministically from there to the start states of each of these machines with a $\lambda, \lambda \to \lambda$ transition, as we did in Figure 5–10.

Key Terms

pushdown automaton • stack alphabet

Exercises

1. What language does the following (2-state) PDA accept?

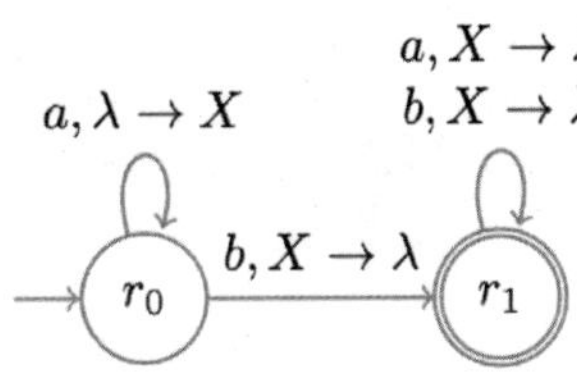

2. Draw a PDA that accepts the language $\{wa^{|w|} \mid w \in (a+b)^*\}$. In other words, an acceptable string is an even-length string where the first half is any string of a's and b's, and the second half is all a's.
3. Draw a PDA that accepts all odd-length strings in $(a+b)^*$ that have a b in the middle.
4. What language does the following PDA accept?

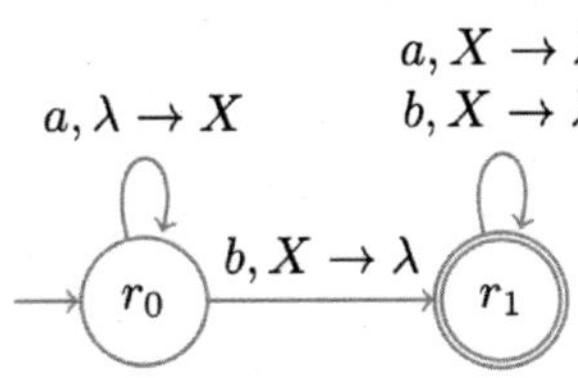

5. What language does the following PDA accept?

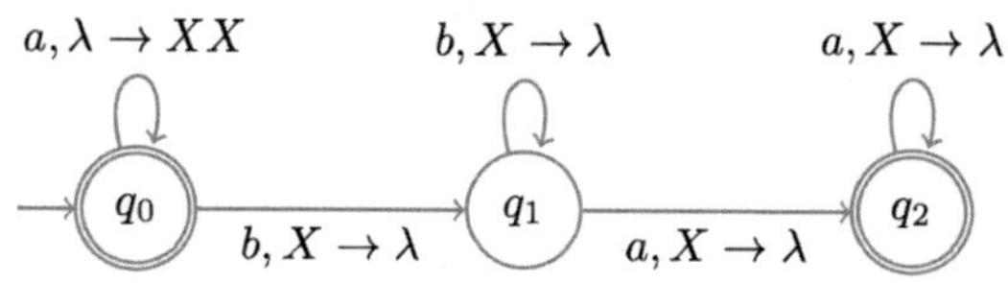

5.2 Pushdown Automata and Determinism

In Figure 5–8 we "seeded" the stack with an X to increase by 1 the number of b's that would be in a string accepted by the PDA. Using such a stack "start symbol" can be useful for other purposes as well. The following PDA for L_{eq} uses a start symbol, S, to know when the stack has "hit bottom" (i.e., when S is at the *top* of the stack, it is the *only symbol* on the stack). Furthermore, whenever an S is on the top of the stack in this machine, it means that an equal number of a's and b's have been encountered so far up to that point.

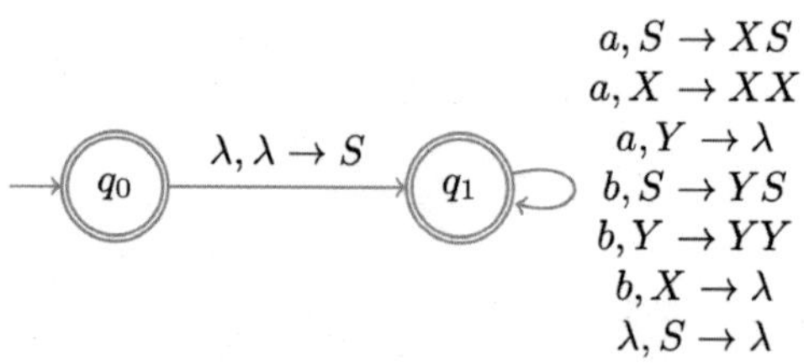

Figure 5–13: Accepts L_{eq} using a stack start symbol/bottom marker

Compare the transitions in state q_1 above with those from Figure 5–8. The following transitions have replaced $a, \lambda \to X$ from Figure 5–8:

$$a, S \to XS$$
$$a, X \to XX$$

Since S is always on the bottom of the stack in this machine, we know that when it is at the top of the stack, nothing else is on the stack, so we can push an X to account for the symbol a just read. The transition $a, Y \to XY$ is omitted because, as before, X's and Y's do not appear on the stack simultaneously in this PDA. The two replacement transitions mentioned above are equivalent to $a, \lambda \to X$ in that their net effect is to add an X to the stack when reading an a.

In like fashion, we replaced the rule $b, \lambda \to Y$ from Figure 5–8 with the following two transitions in Figure 5–13, because their net effect is to push a Y when a b is read:

$$b, S \to YS$$
$$b, Y \to YY$$

and we do not add the rule $b, X \to YX$, since that is not a valid movement in this machine (X's and Y's cannot be on the stack at the same time).

Since we arbitrarily pushed an S on the stack in the beginning, we must add the transition $\lambda, S \to \lambda$ to remove it, so we can end with an empty stack when the input has been exhausted. This is the *only* non-deterministic feature of this PDA (Figure 5–13). All other transitions can be applied deterministically, and we must only apply the $\lambda, S \to \lambda$ transition after reading the last input symbol. Since we replaced the transitions from Figure 5–8 that had lambdas in the pop-position, we have removed some of the non-determinism in the new PDA compared to the original one-state PDA. Hence, using a start symbol as a stack bottom marker can reduce the non-determinism in a PDA. If we knew where the end of the input string was, by using a special symbol, \$, say, we could remove *all* non-determinism (for this particular PDA—it doesn't always work). See Figure 5–14 below.

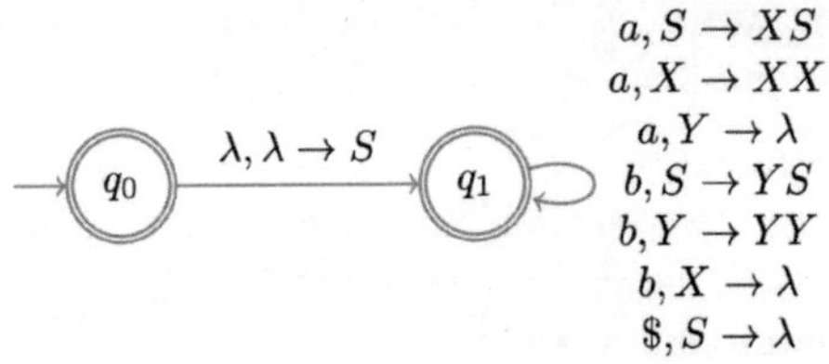

Figure 5–14: Accepts L_{eq} with an end-of-string marker

By using the dollar sign as a special end-of-string marker in the input, if there is an S on top of the stack when reading \$, we know we can pop the S to empty the stack and accept the string. Processing an input string in this fashion makes this a deterministic PDA. The following table traces the processing of the string *abbaba* through this machine.

Table 5–5: Tracing *abbaba*\$

State	Input	Stack
q_0	$abbaba\$$	λ
q_1	$abbaba\$$	S
q_1	$bbaba\$$	XS
q_1	$baba\$$	S
q_1	$aba\$$	YS
q_1	$ba\$$	S
q_1	$a\$$	YS
q_1	\$	S
q_1	λ	λ

We have so far distinguished between deterministic and non-deterministic PDAs, but we haven't yet made clear how to recognize non-determinism in a PDA when we see it. Naturally, if there are multiple transitions for the same combination of state, input symbol, and pop symbol, there is a clear ambiguity. But there are other manifestations of nondeterminism in PDAs.

Let's look a little closer at the mathematical description of the transition function for the language $a^n b^n$ (review Figure 5–1):

$$\delta(q_0, a, \lambda) = (q_0, X)$$
$$\delta(q_0, b, X) = (q_1, \lambda)$$
$$\delta(q_1, b, X) = (q_1, \lambda)$$

This PDA is deterministic because the three input-variable combinations are unique. Suppose that a PDA had the three transitions above along with the additional transition, $\delta(q_0, a, Y) = (q_0, YY)$. This transition *conflicts* with $\delta(q_0, a, \lambda) = (q_0, X)$ because whenever we read an a with a Y on top of the stack, we can either replace the Y with an X or add an additional Y (achieving the latter by popping a Y and then pushing two Y's). The clue that there is a conflict in this case is the presence of a both a *lambda* and a *non-lambda* (Y) pop symbol option whenever we read an a in state q_0.

Now consider adding the transition $\delta(q_1, \lambda, X) = (q_1, Y)$ to the three original transitions above. A similar problem arises with this new transition and the original transition $\delta(q_1, b, X) = (q_1, \lambda)$. The state and pop symbol are identical, but we allow either a lambda or a b for the input symbol, giving us the option of reading an input symbol or not.

So, we can spot non-determinism when multiple transitions from the same state have a lambda and a non-lambda entry for either both pop symbols simultaneously *when reading the same symbol* (including lambda), *or* if they read a symbol and a lambda simultaneously *when popping the same stack symbol* (including lambda). To be deterministic, therefore, if $\delta(q, \lambda, X)$ is defined for any q and X, then *no other transition rule* must exist for that q and X. Likewise, if $\delta(q, c, \lambda)$ is defined for any q and c, then no other rule must exist for that q and c.

There are multiple types of conflicts, made evident by looking at all possible pairings of the types of transitions in the following definition from Sipser[2]:

Definition 5.2

A PDA is **deterministic** if for every $q \in Q$, every $c \in \Sigma$, and every $X \in \Gamma$, *exactly one* of the following transitions is defined:

- $\delta(q, c, X)$
- $\delta(q, c, \lambda)$
- $\delta(q, \lambda, X)$

[2]Sipser, Michael, *Introduction to the Theory of Computation*, Third Edition, CENGAGE Learning, 2013, p. 130.

- $\delta(q, \lambda, \lambda)$

The PDA in Figure 5–13 is not deterministic because $\delta(q_1, \lambda, S)$ conflicts with $\delta(q_1, a, S)$. Note also that $\delta(q, \lambda, \lambda)$ would conflict with *any other transition* on any state q.

If $\delta(q, \lambda, \lambda)$ is defined for any state, q, then any other transition emanating from state q renders the PDA *non-deterministic.*

Definition 5.2 says "exactly one" of the four transitions must be defined, but remember that we typically omit from our diagrams those edges that do not lead to acceptance; they "exist", but may not always be pictured. If we want to "prove" that a PDA is deterministic, we may have to add required edges and an accompanying state to obtain all combinations of valid q, c, and X. See the next example.

Example 5–8

In the diagram below, we add a jail state, q_2, and fill in all missing edges for the PDA for the language L_{eq} in Figure 5–14 to conform to Definition 5.2, giving a DPDA.

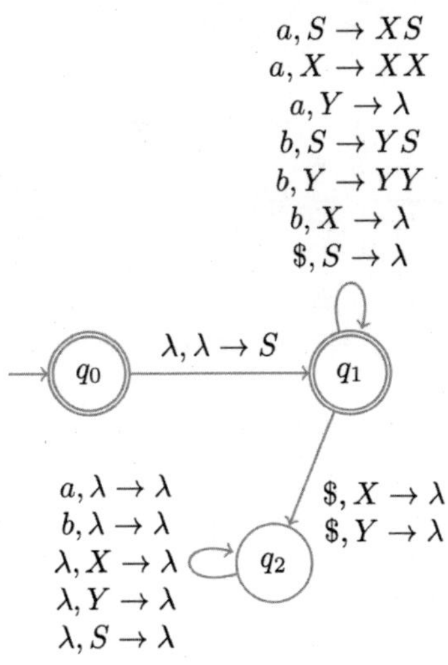

Figure 5–15: Complete DPDA for $n_a(w) = n_b(w)$

You should verify that you understand how this DPDA fulfills Definition 5–2.

Any string other than one of the form $a^n b^n$ will encounter the dollar sign in q_1 with an X or Y on top of the stack instead of an S, so it will transition to state q_2, empty the stack, and reject the string. Since we have emptied the stack in state q_2, we can now form a DPDA for the complement of $n_a(w) = n_b(w)$ by *inverting the acceptability* of each state. DPDAs, like DFAs, are closed under complementation. As with finite automata, we can't obtain the complement of a NPDA by reversing state acceptability—determinism is required.

Deterministic Context-Free Languages (DCFLs) are *closed under complement.* Non-deterministic CFLs (NCFLs) are *not* closed under complement.

Since non-determinism is required to form the union of PDAs, DCFLs are not closed under union, as mentioned previously. (*Determinism is lost* when offering a *choice* via lambda transitions between two DPDAs).

DCFLs are *not* closed under union. NCFLs *are* closed under union.

Key Terms

deterministic pushdown automaton • stack start symbol • end-of-string delimiter

Exercise

1. Find a DPDA for the *complement* of the language $\{a^n b^{2n} \mid n \geq 0\}$. (*Hint*: Make the PDA in Figure 5–2 deterministic by having it conform to Definition 5.2 by adding a stack start symbol (S) and use an end-of-string delimiter ($). Add states and edges as needed. Make sure you empty the stack even for strings not in the language in preparation for forming the complement. You don't have to include processing for impossible cases, however; e.g., you may assume that no input symbols appear after the ending $. Then form the complement of the DPDA. Only states where the stack will be emptied need to be accepting.) Trace the strings *aabbb* and *aabbbbb* to show that they are accepted by your complement. (*Note*: There will be more states than we've seen in most of the previous examples.)

Chapter Summary

Adding a stack as an auxiliary data store to a finite state machine allows recognition of a larger class of languages than without the stack. These are the *context-free languages.* Not all context-free languages are recognized by a *deterministic* PDA, however; some CFLs require non-determinism. Using a *stack start symbol* can help reduce, or in some cases, in conjunction with an *end-of-string symbol,* eliminate the non-determinism in a PDA. The closure properties of NCFLs and DCFLs differ as well, as will be seen in Chapter 7.

6. Context-Free Grammars

Yes, English can be weird. It can be understood through tough thorough thought, though.

– David Burge

Old Rye Grammar School, Simon Harriyott Licensed under CC-BY 2.0

Where Are We?

Language Class	Recognized By	Generated By
Regular	Finite Automata	Regular Grammar Regular Expression
Context-free	Pushdown Automata	Context-free Grammar
Recursively Enumerable	Turing Machine	Unrestricted Grammar

Chapter Objectives

- *Design Context-Free Grammars for context-free languages*
- *Understand the connection between context-free languages and parse trees*
- *Demonstrate the equivalence of context-free grammars and pushdown automata in describing context-free languages*

The term "grammar" conjures up memories of grammar school, where we learned the rules of our native, natural language. A formal grammar, as we have already seen, is an attempt to codify the structure of languages in the abstract, based on research conducted in the 1950s, most notably by Noam Chomsky. A formal grammar consists of:

- **Non-terminal** symbols (also called **variables**), one of which is the *start symbol*
- **Terminal** symbols (the "letters" of the alphabet in use, Σ)
- A set of **rewrite rules**, whereby one begins with the *start variable*, then substitutes replacements for the variables, and finally obtains a string in the language consisting only of terminal symbols

The rewrite rules are also called *production rules*, or just *productions*, for short. Consider the following, simplistic grammar which generates random sentences.

$$\begin{aligned}
\langle sentence\rangle &\rightarrow \langle noun\text{-}phrase\ \rangle\langle verb\text{-}phrase\rangle \\
\langle noun\text{-}phrase\rangle &\rightarrow the\ \langle noun\rangle \\
\langle verb\text{-}phrase\rangle &\rightarrow \langle verb\rangle\ \langle noun\text{-}phrase\rangle \\
\langle verb\rangle &\rightarrow moves \mid annoys \mid deserves \\
\langle noun\rangle &\rightarrow student \mid cow \mid vegetable \mid worm \mid pot
\end{aligned}$$

Up until now we have used single, capital letters for variables in a grammar. When a variable has more than one letter, we enclose it in angle-brackets so that it stands out in contrast to terminal strings. The following derivation produces the string, "the cow annoys the student":

$$\begin{aligned}
\langle sentence\rangle &\Rightarrow \langle noun\text{-}phrase\rangle\ \langle verb\text{-}phrase\rangle \\
&\Rightarrow the\ \langle noun\rangle\ \langle verb\text{-}phrase\rangle \\
&\Rightarrow the\ cow\ \ \langle verb\text{-}phrase\rangle \\
&\Rightarrow the\ cow\ \langle verb\rangle\ \langle nounphrase\rangle \\
&\Rightarrow the\ cow\ annoys\ the\ \langle noun\rangle \\
&\Rightarrow the\ cow\ annoys\ the\ student
\end{aligned}$$

Each step in a derivation is called a **sentential form**. Grammars enforce *syntax*, not meaning (aka *semantics*), so nonsensical strings can result, such as "the vegetable deserves the student".

The left-hand sides of each rule in the grammar above consist solely of a single non-terminal. Such a grammar is called "context-free" because the non-terminals can be replaced by any of their accompanying right-hand side strings, regardless of the context in which the non-terminals appear.

It is customary to substitute for only one variable at a time, and to substitute for the *leftmost* variable at each step in a derivation. Such a derivation is called a **leftmost derivation**.

The following Python program generates five random sentences using this grammar. Observe how the production rules are implemented by separate functions.

Generates random sentences

```
def sentence():
    return noun_phrase() + ' ' + verb_phrase()
def noun_phrase():
    return 'the ' + noun()
def verb_phrase():
    return verb() + ' ' + noun_phrase()
def verb():
    return random.choice(['moves', 'annoys', 'deserves'])
def noun():
    return random.choice(['student', 'cow', 'vegetable', 'worm', 'pot'])

for i in range(5):
    print(sentence())
```

Sample output from this program appears below:

the pot annoys the vegetable
the vegetable moves the worm
the worm annoys the vegetable
the worm moves the cow
the vegetable deserves the worm

In this chapter, we investigate the workings of context-free grammars and show that they describe the same class of languages as pushdown automata.

6.1 Context-Free Grammars and Derivations

We formally define a context-free grammar (CFG) as follows:

Definition 6.1

A **context-free grammar** is a formal grammar consisting of the following:

- A set of *variables* (i.e., a *non-terminal alphabet*), V, one of which is a *start variable*
- An alphabet, Σ, of *terminal symbols*
- A set of *production rules* of the form $v \to s$, where $v \in V$ and $s \in (V \cup \Sigma)^*$, a string of terminals and/or variables

We begin this section by illustrating CFGs for some of the languages seen in the previous chapter.

Example 6–1

The following CFG generates the language $\{a^n b^n \mid n > 0\}$.

$$S \to aSb \mid ab$$

As a reminder, using the alternation symbol, $|$, allows us to combine all the productions for a given variable on a single line. This grammar can also be expressed as:

$$S \to aSb$$
$$S \to ab$$

The smallest string in the language is ab, which occurs by choosing the second rule above. To generate the language $\{a^n b^n \mid n \geq 0\}$, we would replace ab with the empty string.

Example 6–2

To generate the language $\{a^n b^{2n} \mid n \geq 0\}$, we generate two b's for every a:

$$S \to aSbb \mid \lambda$$

We derive the string $aabbbb$ like this:

$$S \Rightarrow aSbb \Rightarrow aaSbbbb \Rightarrow aabbbb$$

Example 6–3

To design a CFG for the language $\{a^i b^j \mid i \neq j\}$, we begin as we did for the language $a^n b^n$, but then require that at least one more of one of the symbols be generated. See the grammar below.

$$S \to aSb \mid X \mid Y$$
$$X \to aX \mid a$$
$$Y \to Yb \mid b$$

After generating equal numbers of each symbol, we choose to add at least one more by substituting X for S, or one more b by choosing Y (we can't do both). Here is a sample derivation:

$$S \Rightarrow aSb \Rightarrow aaSbb \Rightarrow aaYbb \Rightarrow aaYbbb \Rightarrow aabbb$$

Example 6–4

To generate palindromes with a c in the middle, symbols in symmetric positions about the center of the string must be the same:

$$S \rightarrow aSa \mid bSb \mid c$$

The string is built from the *outside-in*. For example, the string $abbcbba$ derives as follows:

$$S \Rightarrow aSa \Rightarrow abbSbba \Rightarrow abbcbba$$

To accept all palindromes over $\Sigma = \{a, b\}$ (without a c), we replace the third rule with three others:

$$S \rightarrow aSa \mid bSb \mid a \mid b \mid \lambda$$

Example 6–5

How would we design a CFG for the language of *non*-palindromes? First, we should ask, is this even a context-free language? In the previous chapter, we stated that the class of non-deterministic context-free languages is *not* closed under complementation. This does not mean that the complement of a non-deterministic CFL is never context-free; it just means that not all complements of non-deterministic CFLs are context-free.

The trick here is to build strings from the outside-in, allowing for optional matching of outwardly symmetric symbols, but then at some point we break the symmetry to get a non-palindrome. Here is a CFG that does just that:

$$\begin{aligned} S &\rightarrow aSa \mid bSb \mid X \\ X &\rightarrow aYb \mid bYa \\ Y &\rightarrow aY \mid bY \mid \lambda \end{aligned}$$

Derivations must eventually choose the variable X, which forces a pair of symbols in outwardly symmetric positions to *differ*. From there we must choose a rule for the variable Y, which allows any substring in $(a + b)^*$ whatsoever to occur in the middle of the now non-palindrome string. Here are two sample derivations:

$$S \Rightarrow aSa \Rightarrow abSba \Rightarrow abXba \Rightarrow abaYbba \Rightarrow ababba$$

$$S \Rightarrow X \Rightarrow bYa \Rightarrow baYa \Rightarrow babYa \Rightarrow babaYa \Rightarrow bababYa \Rightarrow babbaa$$

Example 6–6

Consider the language $a^n b^n a^m = a^n b^n a^*$. We know the CFG for the first two runs of symbols, $a^n b^n$. The trailing a's are independent from the rest of the string, so we can just concatenate an optional number of a's at the end:

$$\begin{aligned} S &\rightarrow XY \\ X &\rightarrow aXb \mid \lambda \\ Y &\rightarrow aY \mid \lambda \end{aligned}$$

The variable X generates the language $a^n b^n$, and Y generates a^*. The start variable, S, generates the *concatenation* of these two languages by simply concatenating X followed by Y in its right-hand-side rule. We have just shown how to concatenate two CFLs if we know their grammars, showing that (non-deterministic[1]) context-free languages are *closed under concatenation.*

Example 6–7

To design a CFG for the language $L_{eq} = \{w \mid w \in (a+b)^* \wedge n_a(w) = n_b(w)\}$, we observe that for every a generated there must also be an accompanying b, but they can come in either of two orderings, and they don't have to be adjacent. Consider the following CFG.

$$S \rightarrow aSbS \mid bSaS \mid \lambda$$

The first rule introduces the a first, and conversely for the second. The symbol S, being the start symbol, represents a string with an equal number of a's and b's. By inserting S between the a and b and at the end of the replacement strings, we do not constrain how an acceptable string can be constructed. Let's derive *baabaabbab*:

$$S \Rightarrow bSaS \Rightarrow baS...$$

At this point observe that we have obtained the first two symbols of the desired string, which substring itself (ba) has an equal number of a's and b's. This substring now precedes an S, which has the same property, so the resulting concatenation will also have an equal number of a's and b's. We can now derive the next two symbols in similar fashion.

[1]We will show later that deterministic CFLs are *not* closed under concatenation.

$$\ldots baS \Rightarrow baaSbS \Rightarrow baabS \ldots$$

We now have the first four terminal symbols. What should we do next? We don't have a rule that generates two a's at once, but we can generate an additional $aabb$ by using the production $aSbS$ twice in a nested fashion:

$$\ldots baabS \Rightarrow baabaSbS \Rightarrow baabaaSbSbS \Rightarrow baabaabSbS \Rightarrow baabaabbS \ldots$$

Once again that we have a string with an equal number of a's and b's. Now we can finish the last portion by generating the final ab from the final remaining S:

$$\ldots baabaabbS \Rightarrow baabaabbaSbS \Rightarrow baabaabbabS \Rightarrow baabaabbab$$

Observe how a leftmost derivation provides an orderly way of generating strings with CFGs.

Example 6–8

Let's now form a CFG for the language $L_{a2b} = \{w \mid w \in (a+b)^* \wedge n_b(w) = 2n_a(w)\}$; each string has twice as many b's as a's, but can appear in any order. Since two b's must appear for every a, the three possible orderings for introducing these three related symbols are $a \cdots b \cdots b$, $b \cdots a \cdots b$, and $b \cdots b \cdots a$, suggesting the following grammar:

$$S \rightarrow aSbSbS \mid bSaSbS \mid bSbSaS \mid \lambda$$

A derivation of $ababbb$ begins as follows:

$$S \Rightarrow aSbSbS \Rightarrow^* aSbb \ldots$$

giving the first a and the last two b's. We departed from a leftmost derivation for convenience here because we needed the first S to generate the inner bab:

$$\cdots aSbb \Rightarrow abSaSbSbb \Rightarrow^* ababbb$$

letting the remaining S's go to λ.

Example 6–9

The following grammar also generates L_{eq}:

$$S \to SS \mid aSb \mid bSa \mid \lambda$$

The second and third productions introduce a and b in the two possible orders, but only produce strings that begin and end with different symbols. However, strings that begin and end with the same symbol, and have an equal number of a's and b's, can be expressed as the concatenation of two strings with equal numbers of both symbols and begin and end with different symbols[2]. For example, $baaabb$ is the concatenation of ba with $aabb$. This is where the production SS applies. The string $baaabb$ derives as follows:

$$S \Rightarrow SS \Rightarrow bSaS \Rightarrow baS \Rightarrow baaSb \Rightarrow baaaSbb \Rightarrow baaabb$$

Simplifying Grammars

Under certain circumstances, the number of variables in a CFG can be reduced. Consider the variable B in the grammar below.

$$A \to a \mid aaA \mid abBc$$
$$B \to abbA \mid b$$

Since the replacement rules for B do not directly refer to B itself, we can easily substitute its two rules where B appears in the third rule for A:

$$A \to a \mid aaA \mid ab\underline{abbA}c \mid ab\underline{b}c$$

This grammar now has only one variable. This type of back substitution is possible when a rule is *not directly recursive.* A variable is directly recursive if it appears anywhere in any of its right-hand side rules.

When we convert PDAs to CFGs in the Section 6.3, you will notice that some rules generated by the conversion procedure don't make sense. For example, a variable, other than the start variable, might never appear on the right-hand side of any rule. Such a variable is *unreachable* in any derivation, and is therefore *useless.* Try to identify a useless variable in the following grammar.

$$S \to A$$
$$A \to aA \mid \lambda$$
$$B \to bA$$

[2] To see why this is so, consider a string of length $2n$ with an equal number of a's and b's that begins and ends with b. Then there are $2n-2$ characters between the outer b's, with n a's and $n-2$ b's. There must therefore be two consecutive a's somewhere in that internal substring, since having a b between every pair of a's would require $n-1$ b's, and we are one short. At some point, where there are two consecutive a's, we can divide the entire string into two substrings of the form $b \cdots a$ and $a \cdots b$, respectively, each having equal numbers of both letters. Try a few examples to convince yourself that this is the case.

Starting with S, the variable B can never appear in a derivation, so we can just delete that rule:

$$
\begin{aligned}
S &\rightarrow A \\
A &\rightarrow aA \mid \lambda
\end{aligned}
$$

Since S is just another name for A here, we can further simplify this grammar to:

$$S \rightarrow aS \mid \lambda$$

This is a regular grammar for the language a^*.

The following grammar also has a useless variable. Can you see why?

$$
\begin{aligned}
S &\rightarrow aSb \mid A \mid \lambda \\
A &\rightarrow aA
\end{aligned}
$$

The variable A never terminates, and, like jails in finite state machines, can therefore never yield a terminal string.

Eliminate all grammar rules that are **unreachable** or **non-terminating**.

Example 6–10

Simplify the following grammar.

$$
\begin{aligned}
S &\rightarrow aS \mid A \mid C \\
A &\rightarrow a \\
B &\rightarrow aa \\
C &\rightarrow aCb
\end{aligned}
$$

B is unreachable, and C is non-terminating, leaving only the following useful rules:

$$
\begin{aligned}
S &\rightarrow aS \mid A \\
A &\rightarrow a
\end{aligned}
$$

We can now back-substitute A to obtain:

$$S \to aS \mid a$$

which is a regular grammar for $a^+ = aa^* = a^*a$.

Example 6–11

Simplify the following grammar.

$$\begin{aligned} S &\to AB \mid AC \\ A &\to aAb \mid bAa \mid a \\ B &\to bbA \mid aaB \mid AB \\ C &\to abCa \mid aDb \\ D &\to bD \mid aC \end{aligned}$$

To determine which variables are useless, it can sometimes be helpful to use a *reachability graph*:

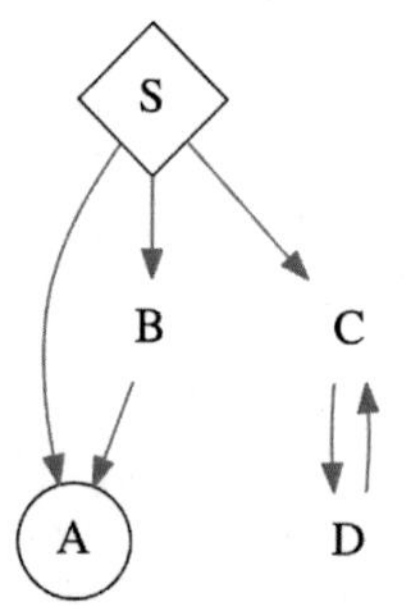

Figure 6–1: Reachability graph for the grammar above

A diamond encloses the start variable. Arrows begin with variables that appear on the left-hand side of a rule, and point to variables that occur in their respective right-hand-side rules. We omit self-loops; we just care where we can reach from the start variable. We enclose variables that have terminal rules in a *circle*. It is obvious by the graph that once a rule containing C is used, there is no way to terminate; we must end at with the variable A. We conclude that both C and D are useless, and remove all rules that contain them. The following grammar emerges:

$$\begin{aligned} S &\to AB \\ A &\to aAb \mid bAa \mid a \\ B &\to bbA \mid aaB \mid AB \end{aligned}$$

Since A and B are both recursive, we can simplify no further.

An unreachable variable is very easy to spot in a reachability graph, as seen by the following graph for the grammar in Example 6–10:

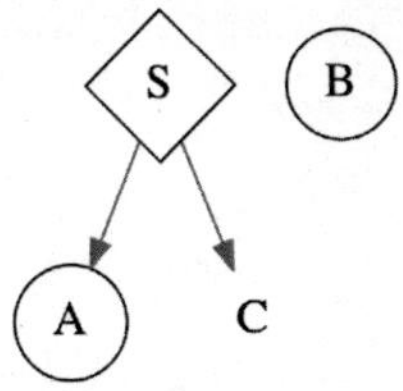

Figure 6–2: Reachability graph for language in Example 6–10

B is a terminating variable, but there is no way to reach it from S.

> **Key Terms**
>
> context-free grammar • non-terminal • derivation • sentential form • useless variable (unreachable, non-terminating) • back-substitution

Exercises

Find context-free grammars for the languages in problems 1–6.

1. $a^n b^{n+2} \mid n \geq 0$.
2. $a^n b^n b^m a^m \mid m, n \geq 0$.
3. $a^n b^m a^m b^n \mid m, n \geq 0$.
4. $w(a+b)^2 w^R \mid w \in (a+b)^*$.
5. $a^n b^m a^q \mid n = m + q$.
6. $a^m b^n a^p b^q \mid m + n = p + q$.
7. Simplify the following CFG:

$$
\begin{aligned}
A &\to 0A1 \mid 0AC \mid 0 \mid 1B0 \\
B &\to 1C0 \mid AC \\
C &\to 1CB \mid AB \\
D &\to 1A0 \mid 1
\end{aligned}
$$

8. Simplify the following CFG:

$$
\begin{aligned}
S &\to A \mid B \\
A &\to \lambda \mid D \\
B &\to aE \\
C &\to \lambda \\
D &\to aDD \\
E &\to aDD \mid aEF \mid bC \\
F &\to bC
\end{aligned}
$$

9. Simplify the following CFG:

$$
\begin{aligned}
S &\to X \mid Y \\
X &\to \lambda \mid aW \mid bB \\
Z &\to \lambda \\
Y &\to aA \mid bC \mid Z \\
W &\to aWW \mid bBW \\
A &\to aAD \mid aWA \mid bBA \mid bCD \mid D \\
B &\to aWB \mid bBB \\
C &\to aBC \mid aAE \mid bBC \mid bCE \mid E \\
D &\to aZ \\
E &\to bZ
\end{aligned}
$$

Programming Exercise

1. Write a program that reads as input an arbitrary context-free grammar and derives strings from the grammar. To illustrate, consider the following grammar for non-palindrome:

$$
\begin{aligned}
S &\to aSa \mid bSb \mid X \\
X &\to aYb \mid bYa \\
Y &\to aY \mid bY \mid \lambda
\end{aligned}
$$

This grammar can be formatted for computer input as follows:

```
S aSa
S aSb
S X
X aYb
X bYa
Y aY
Y bY
Y *
```

Here we are using an *asterisk* for lambda. The first state is assumed to be the start state. You can now read this grammar into a map (aka hash or dictionary) where the key is a variable and the value is a list of right-hand sides. Once you have done this, you can simulate the derivation of a string in the language of a grammar by pushing the start variable onto a stack, and following this procedure:

```
while stack not empty:
    pop a symbol
    if it's a terminal, print it
    otherwise, push one of its right-hand-sides, chosen randomly
```

Generate 10 random strings using the grammar above. Repeat the process on the grammar for $\{a^n b^n \mid n > 0\}$.

6.2 Derivation Trees and Ambiguous Grammars

Derivations of strings in context-free languages have an associated tree representation, called a **derivation tree** (or **parse tree** or **syntax tree**). The advantage of derivation trees is that they reveal the *hierarchical structure* of a derivation. The following diagram shows the tree for the string "the cow annoys the student" using the grammar at the beginning of this chapter. For those of you who diagrammed sentences in elementary school, this will have a somewhat familiar look.

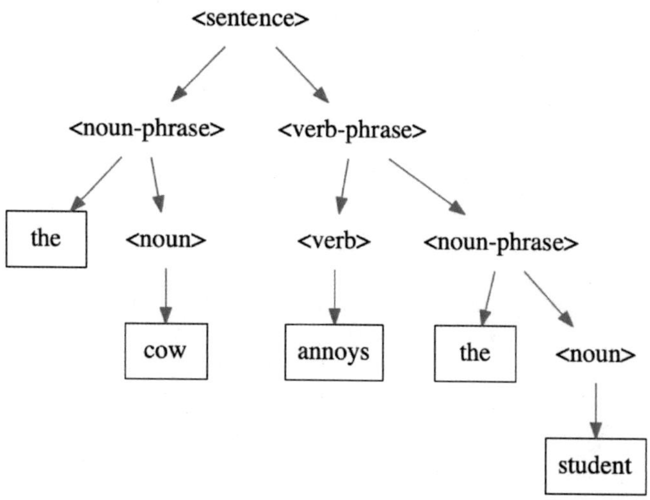

Figure 6–3: Derivation tree for "the cow annoys the student"

The derivation tree clearly depicts the order of substitution in a top-down fashion. The leaves of the tree, enclosed in boxes here for emphasis, are the terminal substrings, which we read from the tree left-to-right to construct the sentence. (We arbitrarily insert spaces between the words for readability.)

Example 6–12

The following grammar forms simple mathematical expressions using addition, multiplication, and parentheses for grouping.

$$\langle exp \rangle \to \langle exp \rangle + \langle exp \rangle \mid \langle exp \rangle * \langle exp \rangle \mid (\langle exp \rangle) \mid a \mid b \mid c$$

Let's derive the expression $a + b * c$ and draw its derivation tree:

$$\langle exp \rangle \Rightarrow \langle exp \rangle + \langle exp \rangle \Rightarrow a + \langle exp \rangle \Rightarrow a + \langle exp \rangle * \langle exp \rangle \Rightarrow a + b * \langle exp \rangle \Rightarrow a + b * c$$

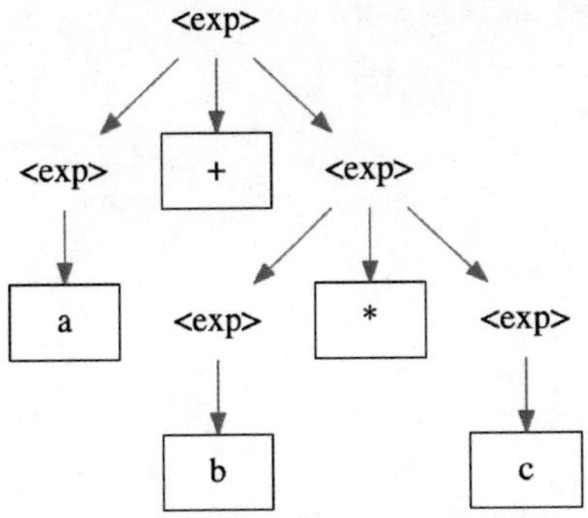

Figure 6–4: Derivation tree for $a + b * c$

It is interesting that with this grammar it is possible to obtain a different leftmost derivation and associated derivation tree for the very same string:

$$\langle exp \rangle \Rightarrow \langle exp \rangle * \langle exp \rangle \Rightarrow \langle exp \rangle + \langle exp \rangle * \langle exp \rangle \Rightarrow a + \langle exp \rangle * \langle exp \rangle \Rightarrow a + b * \langle exp \rangle \Rightarrow a + b * c$$

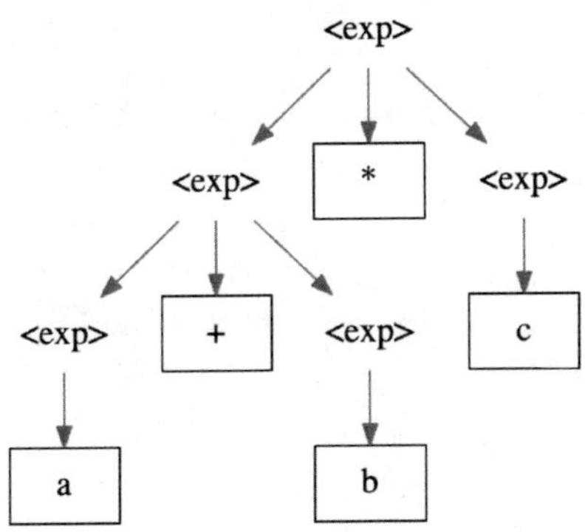

Figure 6–5: Another derivation tree for $a + b * c$

Both trees yield the same sentence, but there is an important structural difference between the two trees. Which one is "correct?"

From a mathematical perspective, the first tree (Figure 6–4) is correct because it represents the correct *order of operations*. The first tree says, "the entire expression is the sum of two subexpressions". What are those two subexpressions? The first subexpression is a and the second is "the product of b and c". Operations further down in the tree evaluate first so that their results are available for the operators higher up in the tree. This is how derivation trees represent **precedence** of operations: the operations *further down* in the tree *evaluate first*.

Operations at the **lower** levels of a derivation tree evaluate **first**.

Operator Precedence

A grammar that gives two distinct derivation trees (or two distinct leftmost derivations) for the *same string* is an **ambiguous grammar**. As with non-determinism, ambiguity is not our friend—it makes implementations difficult. In this case, however, we can modify the grammar to be unambiguous by reflecting the order of operations in the grammar. The usual precedence in mathematics is:

1. expressions grouped by parentheses evaluate first
2. then exponents
3. then multiplication and division
4. then addition and subtraction

Therefore, we want the higher-precedence operations to occur further down in derivation trees, so we will represent them by variables "further down" in the grammar. See the revised grammar below.

$$\begin{aligned}\langle exp\rangle &\to \langle exp\rangle + \langle mulexp\rangle \mid \langle mulexp\rangle\\ \langle mulexp\rangle &\to \langle mulexp\rangle * \langle rootexp\rangle \mid \langle rootexp\rangle\\ \langle rootexp\rangle &\to (\langle exp\rangle) \mid a \mid b \mid c\end{aligned}$$

Addition is at the top of the grammar, so it will appear at the top of a derivation tree, unless it is inside parentheses. Let's now derive the expression $a + b * c$ with the revised grammar:

$$\begin{aligned}\langle exp\rangle &\Rightarrow \langle exp\rangle + \langle mulexp\rangle\\ &\Rightarrow \langle mulexp\rangle + \langle mulexp\rangle\\ &\Rightarrow \langle rootexp\rangle + \langle mulexp\rangle\\ &\Rightarrow a + \langle mulexp\rangle\\ &\Rightarrow a + \langle mulexp\rangle * \langle rootexp\rangle\\ &\Rightarrow a + \langle rootexp\rangle * \langle rootexp\rangle\\ &\Rightarrow a + b * \langle rootexp\rangle\\ &\Rightarrow a + b * c\end{aligned}$$

Note how we "chain" from one level of precedence to the next using the variables. There is no other way to derive the string $a + b * c$ with this grammar, so the derivation tree is unique:

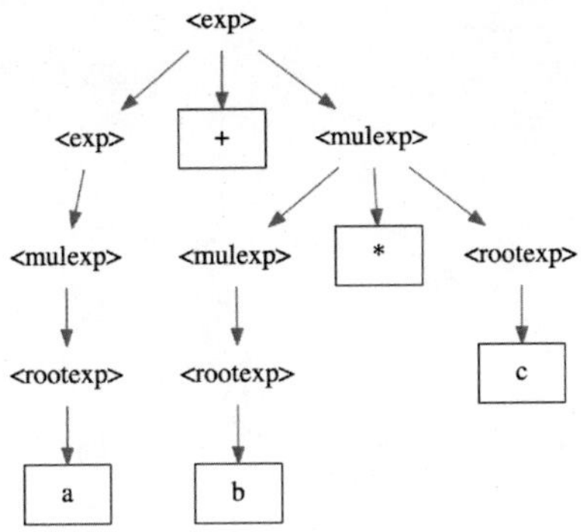

Figure 6-6: Derivation of $a + b * c$ with the revised, unambiguous grammar

Operator Associativity

Let's now construct the derivation tree for the string $a + b + c$ with the revised grammar used above.

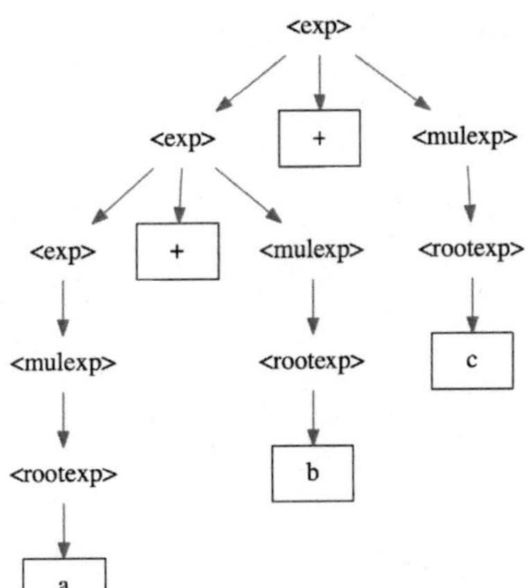

Figure 6-7: Derivation tree for $a + b + c$

The sum $a + b$ is further down in the tree than the second addition, so the order of operations is $(a + b) + c$, which we expect with programming languages. What made this happen was the *left-recursion* in the first rule for the variable $\langle exp \rangle$, which is where the addition operator appears (i.e., the variable on the left-hand side, $\langle exp \rangle$, appears to the *left* of the addition *operator* in the substitution rule on the right). The same goes for $\langle mulexp \rangle$ and the multiplication operator as well.

Example 6–13

In a similar manner, *right-recursion* causes *right-to-left associativity.* To illustrate, below we add an *exponentiation operator* (`^`) to our working grammar.

$$
\begin{aligned}
\langle exp\rangle &\to \langle exp\rangle + \langle mulexp\rangle \mid \langle mulexp\rangle\\
\langle mulexp\rangle &\to \langle mulexp\rangle * \langle powerexp\rangle \mid \langle powerexp\rangle\\
\langle powerexp\rangle &\to \langle rootexp\rangle\hat{}\langle powerexp\rangle \mid \langle rootexp\rangle\\
\langle rootexp\rangle &\to (\langle exp\rangle) \mid a \mid b \mid c
\end{aligned}
$$

The rule for $\langle powerexp\rangle$ is *right-recursive.* Here is the derivation tree for the string `a+b*c^d^(e+f)` which represents the expression $a + b * c^{d^{e+f}}$:

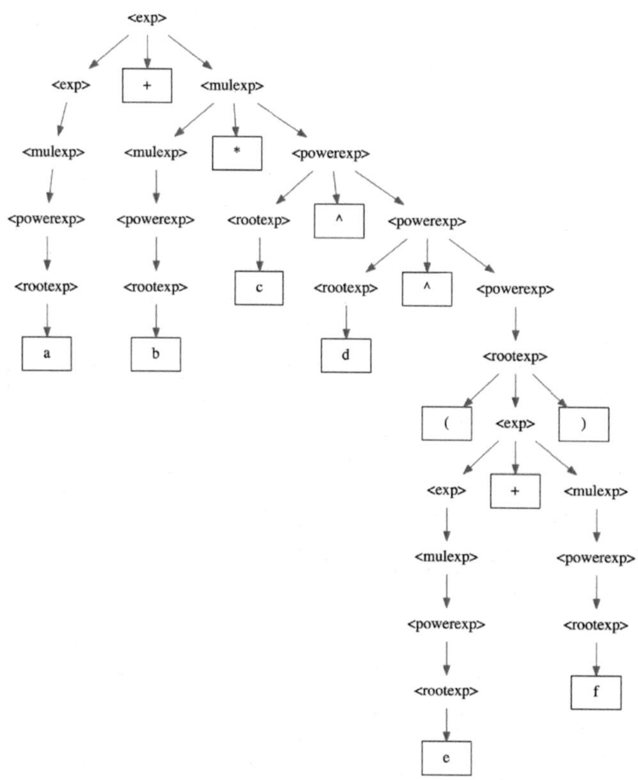

Figure 6–8: Derivation tree for the string `a+b*c^d^(e+f)`

Observe that the exponentiation operator is *right associative* because the variable $\langle powerexp\rangle$ is *right-recursive* with respect to its operator.

For left-to-right associativity, use left recursion in the variable for the operation. For right-to-left associativity, use right recursion.

Expression Trees

The tree in Figure 6–8 is complex and nearly fills a page. For convenience, parsers in compilers typically collapse all chains of single-variable substitutions, leaving only leaves of terminals in the tree. The result, called an **expression tree**, is easier to process programmatically and clearer to grasp visually. See the following diagram.

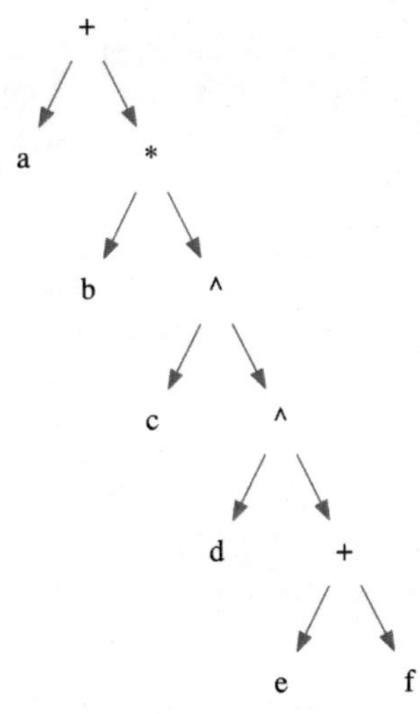

Figure 6–9: Expression tree for Figure 6–8

Since everything is a leaf, we omit the boxes. We also omit the parentheses, since the sum $e + f$ was pushed further down in the tree by the precedence of parentheses in the grammar. This tree makes clear that the entire expression is the sum of a and a product. That product is b multiplied by an exponentiation. That exponentiation is c raised to another exponentiation, namely d^{e+f}.

One technique to obtain an expression from a derivation tree is to move all terminals up, bypassing all non-terminals, row-by-row in a top-down fashion. For example, the boxed $+$ in row 2 of the derivation tree in Figure 6–7 can ascend to replace the $\langle exp \rangle$ at the top (row 1). The boxed $*$ in row 3 can ascend to replace $\langle mulexp \rangle$ in row 2. The a in row 6 can ascend to replace the $\langle exp \rangle$ in row 2.

It is also possible to build an expression tree for an expression directly, bypassing the grammar altogether, simply by knowing the order of operations. We can easily see that the last operation for the expression $a + b * c^{d^{e+f}}$ is the addition with a on the left, so we place that $+$ at the top of the tree with a as its left child node. The next-to-last operation is the multiplication with b on the left, etc. It is also possible to construct the tree bottom-up starting with the first operation the computer will perform.

Mathematical expressions can be represented in text without any parentheses whatsoever in a way that maintains the order of operations. The two common techniques are **prefix** notation and **postfix** notation. These notations require special algorithms for evaluating them.

Prefix notation can be formed from an expression tree by traversing the tree counter-clockwise, starting at the root, and collecting each node's string left-to-right the first time you encounter them (this is a *preorder traversal* of the tree). Traversing the tree in Figure 6–9 in preorder yields the following string (appearing here with a space between each token for readability):

```
+ a * b ^ c ^ d + e f
```

To evaluate this expression, we scan right-to-left looking for an operator. Apply the operator to its subsequent two operands (`e` and `f`, in this case). Replace all three tokens with the result of applying the operator to the operands and repeat the process. To make this example concrete, we will assign values to the variables as follows.

```
+ 3 * 2 ^ 2 ^ + 2 1
```

To begin the evaluation of the prefix expression, we scan from right-to-left for an operator, which happens to be the rightmost `+`, so we replace `+ 2 1` with `3`:

```
+ 3 * 2 ^ 2 ^ 2 3
```

Repeating the process, we replace `^ 2 3` with `8`:

```
+ 3 * 2 ^ 2 8
```

and then `^ 2 8` with `256`, and so on:

```
+ 3 * 2 256
+ 3 512
515
```

To express the same computation in *postfix notation*, we traverse the tree in Figure 6–9 clockwise, and then reverse the resulting string (which is equivalent to a *postorder* traversal). Traversing clockwise gives the string:

```
+ * ^ ^ + f e d c b a
```

which, when reversed, becomes:

```
a b c d e f + ^ ^ * +
```

To evaluate this postfix expression, we traverse left-to-right looking for an operator, applying it to the two operands that precede it, and replacing the three tokens with the result of the operation. Using the same values as before, we perform the same computations in the same order:

```
3 2 2 2 2 1 + ^ ^ * +
3 2 2 2 3 ^ ^ * +
3 2 2 8 ^ * +
3 2 256 * +
3 512 +
515
```

The following Python code implements both traversals of the tree in Figure 6–9.

Prefix/Preorder and Postfix/Postorder traversals

```
''' Output:
Preorder: + a * b ^ c ^ d + e f
Postorder:  a b c d e f + ^ ^ * +
'''

class Node:
    def __init__(self, label, left=None, right=None):
        self.label = label
        self.left = left
        self.right = right

def preorder(node):
    if node:
        return node.label + ' ' + preorder(node.left) + preorder(node.right)
    else:
        return ''

def postorder(node):
    if node:
        return postorder(node.left) + postorder(node.right) + ' ' + node.label
    else:
        return ''
```

```
expr = Node('+',
    Node('a'), Node('*',
        Node('b'), Node('^',
            Node('c'), Node('^',
                Node('d'), Node('+',
                    Node('e'), Node('f'))))))
print('Preorder:', preorder(expr))
print('Postorder:', postorder(expr))
```

Key Terms

derivation tree • operator precedence • operator associativity • left recursion • right recursion • prefix notation • postfix notation

Exercises

1. What language is generated by the following CFG?

$$
\begin{aligned}
S &\rightarrow aSb \mid X \\
X &\rightarrow cXd \mid cYd \\
Y &\rightarrow aYb \mid ab
\end{aligned}
$$

2. Derive the string *aacabdbb* using the grammar in the previous exercise.
3. Draw the derivation tree for the derivation in the previous exercise.
4. Show that the following grammar for balanced parentheses is ambiguous.

$$S \rightarrow SS \mid (S) \mid \lambda$$

5. Modify the grammar below to add rules for subtraction (–), division (/), bitwise-or (|) and bitwise-and (&). The bitwise operations have the same precedence and are the lowest precedence operators. Subtraction is the same precedence as addition, and division is the same precedence as multiplication. You may want to put quotes around the "|" for the bitwise-or to distinguish it from the normal alternation symbol.

$$E \to E + M \mid M$$
$$M \to M * R \mid R$$
$$R \to (E) \mid a \mid b \mid c$$

6. Draw a derivation tree using your answer in the previous exercise for the string `(a|b) / (b&c) * (a-c)`. Make sure the last operation to execute at runtime is the first operator that appears at the top of the tree.
7. Draw the expression tree for your derivation tree in the previous problem.
8. Using your result from the previous problem, write the expression in *postfix* notation.
9. Write the same expression as above in *prefix* notation.

6.3 Equivalence of PDAs and CFGs

We now show that pushdown automata and context-free grammars describe the same class of languages. As we did with regular expressions and finite automata, we construct one from the other in a way that preserves the language in question. One way is trivial and the other is very involved. We start with the easier case first.

From CFG to PDA

A non-deterministic pushdown automaton can easily simulate the rules of a context-free grammar as follows:

1. The PDA has two states: a non-accepting start state and a second, final state.
2. Place a single transition from the start state to the final state that simply pushes the CFG's start variable on the stack.
3. For every grammar rule, $v \to r$, where v is a variable in the grammar and r is v's replacement, place a self-loop transition on the final state that consumes no input, pops v, and pushes r—that is, the transition $\lambda, v \to r$.
4. For every terminal symbol, c, in the alphabet, Σ, add a self-loop transition on the final state that reads and pops c—that is, the transition $c, c \to \lambda$.

We illustrate the process by creating a PDA for the grammar for the language L_{eq} from Example 6–9. See the diagram below.

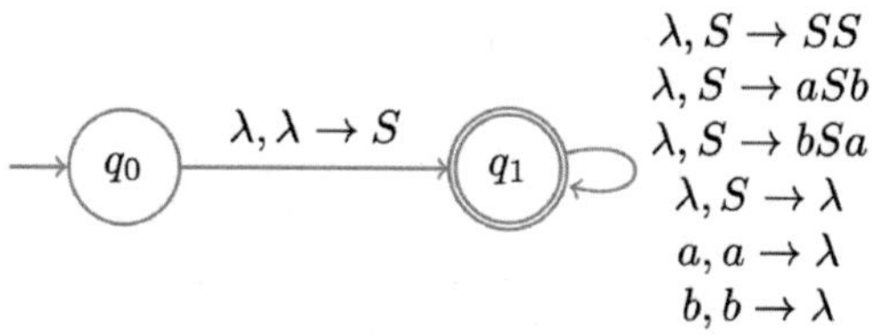

Figure 6–10: PDA simulating the rules of the grammar in Example 6–9

This PDA processes a string by applying the transitions corresponding to the rules used in Example 6–9. Whenever a symbol of the alphabet is on the top of the stack, there will be a symbol in the input to match, if the input string is valid and we have applied the rules correctly, at which point we apply the rule that reads and pops that symbol. The table below processes the same string, *baaabb*, from Example 6–9.

Table 6–1: Simulates the derivation of *baaabb* in Example 6–9

State	Input	Stack	Transition to Apply Next
q_0	$baaabb$	λ	$\lambda, \lambda \to S$
q_1	$baaabb$	S	$\lambda, S \to SS$
q_1	$baaabb$	SS	$\lambda, S \to bSa$
q_1	$baaabb$	$bSaS$	$b, b \to \lambda$
q_1	$aaabb$	SaS	$\lambda, S \to \lambda$
q_1	$aaabb$	aS	$a, a \to \lambda$
q_1	$aabb$	S	$\lambda, S \to aSb$
q_1	$aabb$	aSb	$a, a \to \lambda$
q_1	abb	Sb	$\lambda, S \to aSb$
q_1	abb	$aSbb$	$a, a \to \lambda$
q_1	bb	Sbb	$\lambda, S \to \lambda$
q_1	bb	bb	$b, b \to \lambda$
q_1	b	b	$b, b \to \lambda$
q_1	λ	λ	(accept)

Example 6–14

The following PDA simulates the grammar for palindromes from Example 6–4.

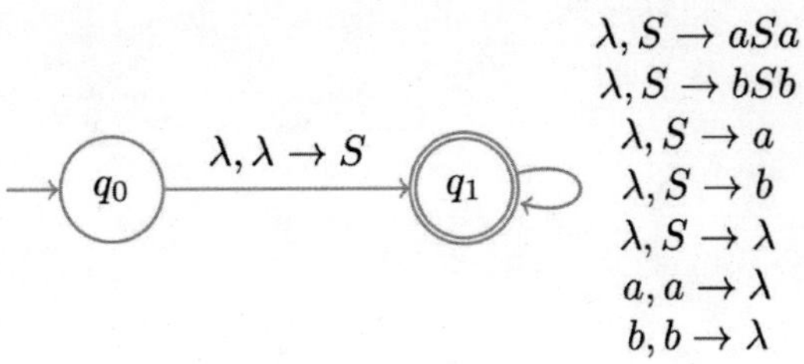

Figure 6–11: PDA simulating the rules of the palindrome grammar

The following table simulates the derivation of $abbbba$.

Table 6–2: Accepts $abbbba$

State	Input	Stack	Transition to Apply Next
q_0	$abbbba$	λ	$\lambda, \lambda \to S$
q_1	$abbbba$	S	$\lambda, S \to aSa$
q_1	$abbbba$	aSa	$a, a \to \lambda$
q_1	$bbbba$	Sa	$\lambda, S \to bSb$
q_1	$bbbba$	$bSba$	$b, b \to \lambda$
q_1	$bbba$	Sba	$\lambda, S \to bSb$
q_1	$bbba$	$bSbba$	$b, b \to \lambda$
q_1	bba	$Sbba$	$\lambda, S \to \lambda$
q_1	bba	bba	$b, b \to \lambda$
q_1	ba	ba	$b, b \to \lambda$
q_1	a	a	$a, a \to \lambda$
q_1	λ	λ	(accept)

From PDA to CFG (Special Case)

The other direction (PDA→ CFG) is much more tedious, except that there is a special case that is simple: a **single-state PDA**. The language L_{eq} in Example 5–4 is recognized by such a PDA (repeated in the diagram below).

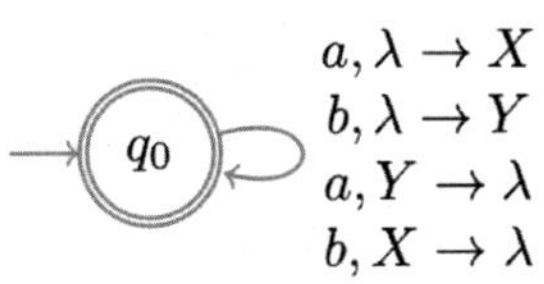

Copy of Figure 5–7

To convert a single-state PDA to a CFG we do the following:

1. Add a new, non-accepting start state that pushes a unique variable, S, say, in a transition that moves to the original, single accepting state. This will be the grammar's start variable.
2. Add the transition $\lambda, S \to \lambda$ to the original (now second) state to compensate for having added S to the stack, thus allowing it to be popped so the stack can be emptied.
3. For every transition on the original state that **pops lambda**, $c, \lambda \to r$, say, *replace* it with equivalent transitions of the form $c, V \to rV$ for **every stack symbol** $V \in \Gamma$ (including S).
4. To form the CFG, for each transition $c, V \to r$ on the final state add the grammar rule $V \to cr$. S is the start variable.

The following PDA reflects the changes in steps 1–2 above for the PDA for L_{eq} in Figure 5–7:

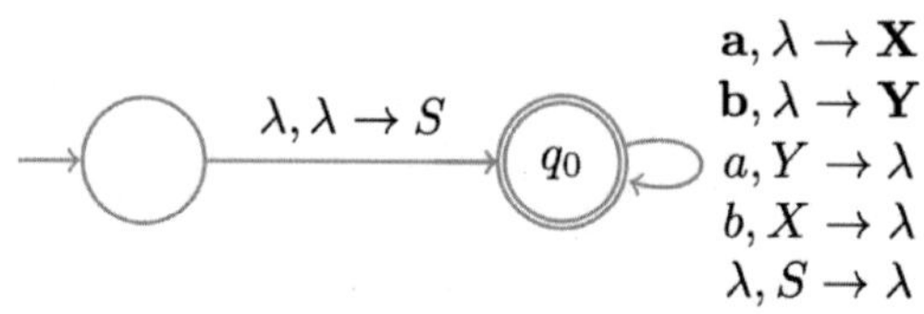

Figure 6–12: Modifying the single-state PDA for L_{eq} for conversion to a CFG

The first two transitions on the final state pop lambda, so we replace them according to step 3 above, noting that we now have three stack variables (S, X, Y):

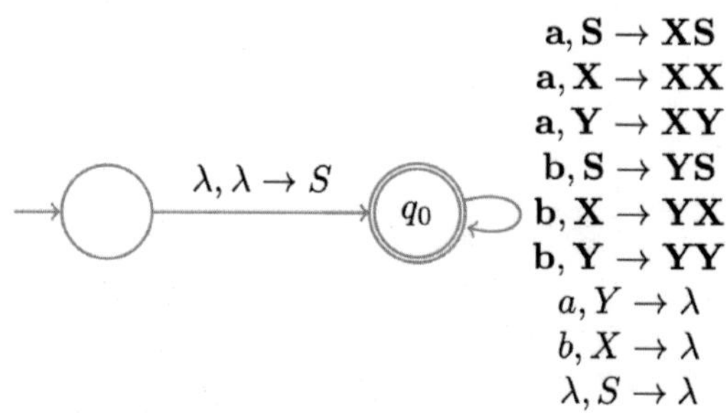

Figure 6–13: Inserting the replacement transitions for lambda-pops

We can now follow step 4 above and form the grammar directly from the transitions on state q_0. Note that the consumed stack variable is replaced by the consumed symbol concatenated with the pushed stack elements. These become our grammar rules. Below we group rules by each stack variable:

$$S \to aXS \mid bYS \mid \lambda$$
$$X \to aXX \mid bYX \mid b$$
$$Y \to aXY \mid bYY \mid a$$

If you look back at the original one-state PDA in Example 5–4 in the previous chapter, you will recall that we omitted the transitions $a, Y \to XY$ and $b, X \to YX$. That was a special case for this example only, because the PDA was not designed to apply those actions (recall that X and Y never appear simultaneously on the stack). The corresponding rules bYX and aXY can also be omitted here. You can always omit needless rules on a case-by-case basis.

Example 6–15

The following one-state PDA accepts strings of balanced parentheses:

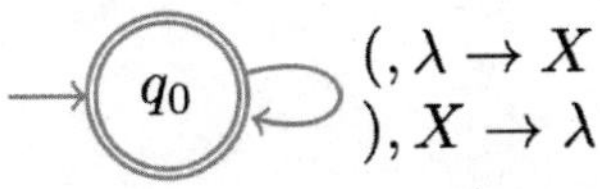

Figure 6–14: One-state PDA accepting balanced parentheses

To find a CFG for this PDA, we add the new start state, add transitions to push and then pop S, and replace the lambda-pop transition with equivalent, non-lambda-pop transitions:

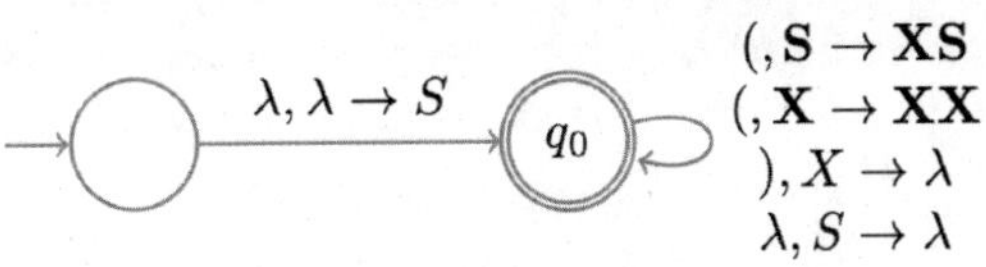

Figure 6–15: Modified PDA to extract CFG

We can now extract the grammar:

$$S \to (XS \mid \lambda$$
$$X \to (XX \mid)$$

To verify that we get a valid string, we derive $(()())$:

$$S \Rightarrow (XS \Rightarrow ((XXS \Rightarrow (()XS \Rightarrow (()(XXS \Rightarrow (()()XS \Rightarrow (()())S \Rightarrow (()())$$

From PDA to CFG (General Case)

To convert an arbitrary PDA to a CFG, we somehow need to relate the **states and transitions** of the PDA, including **stack operations**, to **variables and rules** in a CFG. This is a tall order, but a doable one. To understand the algorithm, we need to take a deeper look at stack behavior in PDAs. We start with the simple, one-state PDA for L_{eq} found in Figure 5–7. While referring to that figure, study the following trace of $aaababbbbbaa$.

Table 6–3: Trace of $aaababbbbbaa$

Step	Input	Stack
0	$aaababbbbbaa$	λ_1
1	$aababbbbbaa$	X_1
2	$ababbbbbaa$	X_2X_1
3	$babbbbbaa$	$X_3X_2X_1$
4	$abbbbbaa$	X_2X_1
5	$bbbbbaa$	$X_4X_2X_1$
6	$bbbbaa$	X_2X_1
7	$bbbaa$	X_1
8	$bbaa$	λ_2
9	baa	Y_1
10	aa	Y_2Y_1
11	a	Y_1
12	λ	λ

We have numbered the stack variables (and the lambdas) with subscripts according to the order in which they appear on the stack so we can inspect their *stack lifetimes.* As always, the stack begins and ends empty, and in between it grows and shrinks in size repeatedly. A stack variable's lifetime is the number of steps from when it is first pushed until it is removed from the stack. The lifetimes of the stack variables in Table 6–3 appear in the following table.

Table 6–4: Stack variable lifetimes from Table 6–3

Stack Variable	Lifetime (in steps)	Step Range
X_1	7	1–7
X_2	5	2–6
X_3	1	3–3
X_4	1	5–5
Y_1	3	9–11
Y_2	1	10–10

From the time the first variable, X_1, appears on the stack until it is removed, three other X's (X_2, X_3, and X_4) come and go. We call these actions the *after-effects* of X_1. The very presence of a stack variable leads to associated after-effects, depending on which transitions subsequently execute. The second X to appear on the stack, X_2, causes the variables X_3 and X_4 to come and go. These actions also *consume input symbols.* Although this PDA has only one state, that is not always the case with an arbitrary PDA. The way we relate the actions of any PDA to grammar rules is by means of the following abstraction.

The symbol $\langle pVq \rangle$ represents the actions of a PDA that move from state p, having the symbol V

at the top of the stack, to state q after *one or more transitions*, such that V and its *after-effects* have all been popped from the stack.

In other words, $\langle pVq \rangle$ represents the one or more symbols consumed while V and its after-effects were on the stack, and we were in state p when V was popped, and we entered state q at the moment V and its after-effects were completely removed. Symbols of this form will become the *variables* in the resulting grammar, representing the sequence of movements in the PDA that reads a substring and pops a stack variable along with all its after-effects, all the while traveling through an arbitrary number of states. In the example above, we have the following correspondences between 3-tuple symbols of the form $\langle pVq \rangle$ and movements in the machine, listed in the order in which they are removed from the stack (review Table 6–3). We also include the configurations starting with an empty stack (corresponding to lines 0 and 8 from Table 6–3 earlier):

Table 6–5: Relating new variable symbols to machine actions

Symbol	PDA Action	String Consumed
$\langle q_0X_3q_0 \rangle$	$(q_0, babbbbbaa, X_3X_2X_1) \vdash (q_0, abbbbbaa, X_2X_1)$	b
$\langle q_0X_4q_0 \rangle$	$(q_0, bbbbbaa, X_4X_2X_1) \vdash (q_0, bbbbaa, X_2X_1)$	b
$\langle q_0X_2q_0 \rangle$	$(q_0, ababbbbbaa, X_2X_1) \vdash^* (q_0, bbbaa, X_1)$	$ababb$
$\langle q_0X_1q_0 \rangle$	$(q_0, aababbbbbaa, X_1) \vdash^* (q_0, bbaa, \lambda)$	$aababbb$
$\langle q_0Y_2q_0 \rangle$	$(q_0, aa, Y_2Y_1) \vdash (q_0, a, Y_1)$	a
$\langle q_0Y_1q_0 \rangle$	$(q_0, baa, Y_1) \vdash^* (q_0, \lambda, \lambda)$	$babb$
$\langle q_0\lambda q_0 \rangle$	$(q_0, bbaa, \lambda_2) \vdash^* (q_0, \lambda, \lambda)$	$bbaa$
$\langle q_0\lambda q_0 \rangle$	$(q_0, aaababbbbbaa, \lambda_1) \vdash^* (q_0, \lambda, \lambda)$	(entire string)

The symbol $\langle q_0X_1q_0 \rangle$, for the first X pushed, represents steps 1–7 in the trace in Table 6–3, which corresponds to consuming the characters $aababbb$ as part of the after-effects of X_1 (starting with the second a in the original string).

The way we relate this to a grammar rule is by noticing that, starting from step 1, the transition $a, \lambda \rightarrow X$ was used to push the second X (X_2). We can express this step of the PDA as the grammar rule

$$\langle q_0\lambda q_0 \rangle \rightarrow a\langle q_0, X, q_0 \rangle$$

meaning that we didn't care what was on the stack (no pop occurred; hence the lambda in the pop position)—we just pushed a new X. Had there been a Y on top of the stack, we would have used a different transition to pop it. The right-hand side of the rule generates an a because the PDA consumed an a, and the variable $a\langle q_0, X, q_0 \rangle$ indicates that we stayed in state q_0, and that subsequent actions are still needed to remove a subsequent X (X_2) and *its* after-effects, eventually ending in state q_0 (the latter due to the trailing q_0 in the variable on the left).

We still have a lot to do to finish the grammar. First, recall that the variable $\langle q_0\lambda q_0 \rangle$ means, "from state q_0, perform one or more PDA actions and end in state q_0, with the stack unchanged from what it was" (that's the effect of having λ in the pop-position). This also happens to represent the overall goal of a PDA—to start in the initial state, and end in a final state, with the stack empty like it was in the beginning. For this reason, our first grammar rule will be:

$$S \rightarrow \langle q_0 \lambda q_0 \rangle$$

In general, we form rules $S \rightarrow \langle s \lambda f \rangle$, where s is the start state, for *all final states, f*.

Next, we have every state go to *itself* without disturbing the stack or generating any symbols:

$$\langle q \lambda q \rangle \rightarrow \lambda$$

for all $q \in Q$. This means we can just stay in the state and do nothing, which doesn't sound terribly useful, but such rules come in handy as terminators, as we'll soon see[3].

The next step is to *add*[4] equivalent transitions for every lambda-pop transition *and* every stack variable, like we did in Figure 6–13 and Figure 6–15 (except we *don't* discard the original lambda-pop rule this time!). We call this **normalizing** the PDA. The table below lists the additional transitions needed.

Table 6–6: Additional transitions to normalize the PDA above

Original Transition	Additional Transitions
$a, \lambda \rightarrow X$	$a, X \rightarrow XX; a, Y \rightarrow XY$
$b, \lambda \rightarrow Y$	$b, X \rightarrow YX; a, Y \rightarrow YY$

The point here is that we must explicitly entertain *all possibilities* of PDA configurations, so that we don't miss any ways that PDA transitions may apply. The normalized PDA for L_{eq} appears in the following diagram.

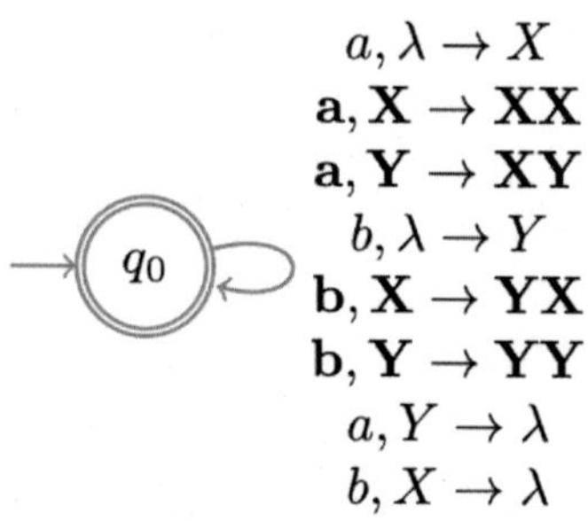

Figure 6–16: Normalized PDA for L_{eq}

We now form grammar rules for each of the of the transitions above, as explained earlier:

[3]We call these "nullity rules".

[4]We *keep* the original lambda-pop rule, instead of replacing it as we did earlier with the special case conversion from a single-state PDA to CFG.

Table 6–7: Grammar rules for transitions

Transition	New Grammar Rule
$a, \lambda \to X$	$\langle q_0 \lambda q_0 \rangle \to a \langle q_0 X q_0 \rangle$
$a, X \to XX$	$\langle q_0 X q_0 \rangle \to a \langle q_0 X q_0 \rangle \langle q_0 X q_0 \rangle$
$a, Y \to XY$	$\langle q_0 Y q_0 \rangle \to a \langle q_0 X q_0 \rangle \langle q_0 Y q_0 \rangle$
$b, \lambda \to Y$	$\langle q_0 \lambda q_0 \rangle \to b \langle q_0 Y q_0 \rangle$
$b, X \to YX$	$\langle q_0 X q_0 \rangle \to b \langle q_0 Y q_0 \rangle \langle q_0 X q_0 \rangle$
$b, Y \to YY$	$\langle q_0 Y q_0 \rangle \to b \langle q_0 Y q_0 \rangle \langle q_0 Y q_0 \rangle$
$a, Y \to \lambda$	$\langle q_0 Y q_0 \rangle \to a \langle q_0 \lambda q_0 \rangle$
$b, Y \to \lambda$	$\langle q_0 X q_0 \rangle \to b \langle q_0 \lambda q_0 \rangle$

The grammar rule must have a 3-tuple variable on the right-hand side of the rule for each stack symbol pushed. For readability, we rename the grammar variables as follows:

$$\begin{aligned} \langle q_0 \lambda q_0 \rangle &\leftrightarrow L \\ \langle q_0 X q_0 \rangle &\leftrightarrow X \\ \langle q_0 Y q_0 \rangle &\leftrightarrow Y \end{aligned}$$

We now group all the rules by the newly renamed variables:

$$\begin{aligned} &S \to L \\ &L \to \lambda \mid aX \mid bY \\ &X \to aXX \mid bYX \mid bL \\ &Y \to aXY \mid bYY \mid aL \end{aligned}$$

S is the same as L here, so we'll just bypass L:

$$\begin{aligned} &S \to \lambda \mid aX \mid bY \\ &X \to aXX \mid bYX \mid bS \\ &Y \to aXY \mid bYY \mid aS \end{aligned}$$

This grammar is similar to others we have already seen for L_{eq}. Here is a derivation for $aaababbbbbaa$:

$$\begin{aligned} S &\Rightarrow aX \Rightarrow aaXX \Rightarrow aaaXXX \Rightarrow aaabSXX \Rightarrow aaabXX \Rightarrow aaabaXXX \\ &\Rightarrow aaababSXX \Rightarrow aaababXX \Rightarrow aaababbSX \Rightarrow aaababbX \Rightarrow aaababbbS \\ &\Rightarrow aaababbbbY \Rightarrow aaababbbbbYY \Rightarrow aaababbbbbaSY \Rightarrow aaababbbbbaY \\ &\Rightarrow aaababbbbbaaS \Rightarrow aaababbbbbaa \end{aligned}$$

Example 6–16

Let's do one more single-state example by converting the PDA in Figure 6–14 to a CFG. First, we write the following automatic (aka "boilerplate") rules mentioned in the previous example, the first representing the overall goal of the PDA, and the second representing the state "nullity" rule:

$$S \to \langle q_0 \lambda q_0 \rangle$$
$$\langle q_0 \lambda q_0 \rangle \to \lambda$$

Next, we normalize the PDA:

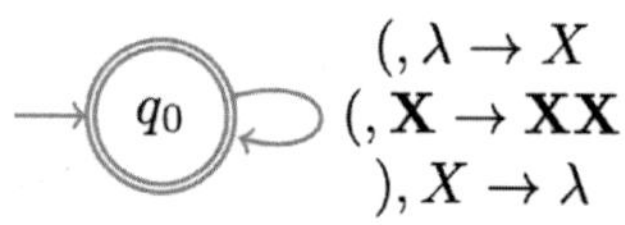

Figure 6–17: Normalizing the PDA in Figure 6–14

Now we convert each transition to a grammar rule:

$$\langle q_0 \lambda q_0 \rangle \to (\langle q_0 X q_0 \rangle$$
$$\langle q_0 X q_0 \rangle \to (\langle q_0 X q_0 \rangle \langle q_0 X q_0 \rangle$$
$$\langle q_0 X q_0 \rangle \to)\langle q_0 \lambda q_0 \rangle$$

After renaming $\langle q_0 \lambda q_0 \rangle$ as L and $\langle q_0 X q_0 \rangle$ as X, we gather the rules on a single line for each variable:

$$S \to L$$
$$L \to \lambda \mid (X$$
$$X \to (XX \mid)L$$

Once again, we bypass L giving:

$$S \to \lambda \mid (X$$
$$X \to (XX \mid)S$$

As a sanity check we derive ()(()()):

$$S \Rightarrow (X \Rightarrow ()S \Rightarrow ()(X \Rightarrow ()((XX \Rightarrow ()(()SX \Rightarrow ()(()X$$
$$\Rightarrow ()(()(XX \Rightarrow ()(()()SX \Rightarrow ()(()()X \Rightarrow ()(()())S \Rightarrow ()(()())$$

Example 6–17

The following PDA that accepts $\{a^nb^n \mid n \geq 0\}$:

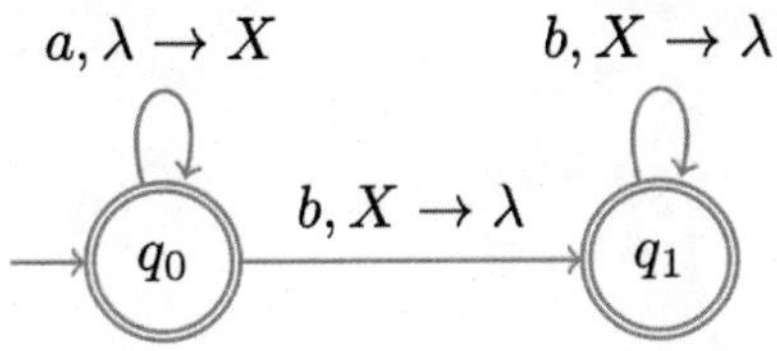

Figure 6–18: PDA accepting $\{a^nb^n \mid n \geq 0\}$

The convert this PDA to a CFG, we first normalize it by adding the transition $a, X \to XX$ on state q_0:

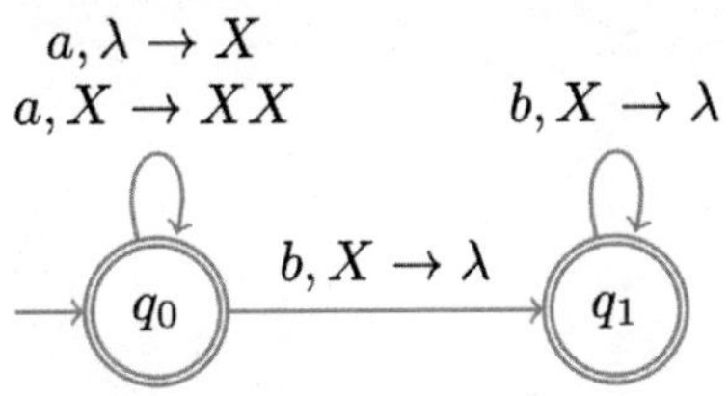

Figure 6–19: Figure 6–18 normalized

There are two accepting states, so the first boilerplate rules are:

$$S \to \langle q_0 \lambda q_0 \rangle \mid \langle q_0 \lambda q_1 \rangle$$

As usual, we have nullity rules for each state:

$$\langle q_0 \lambda q_0 \rangle \to \lambda$$
$$\langle q_1 \lambda q_1 \rangle \to \lambda$$

With multiple states, we must entertain *all possible combinations* of states as we convert each transition to grammar rules. This is where things get tedious. Consider the first transition, $a, \lambda \to X$. We begin to extract a grammar rule for this transition as follows:

$$\langle q_0 \lambda - \rangle \to a \langle q_0 X - \rangle$$

We did not yet populate the third slot (marked by a dash) of each grammar variable yet. In the previous examples there was only one state to consider, but now there are two states. Recall that the interpretation of the incomplete rule above is, "starting in state q_0, we pop nothing, consume

a, push X, and move in *one step* to state q_0." Where we eventually end up after multiple moves is reflected in the third slot of each variable, and by the definition of the notation $\langle pVq \rangle$, those slots must represent the *same state*—consuming a, pushing X, and staying in q_0 was just the first step in that journey. Since we must consider all possibilities of target states, we form two rules, one for each state in the machine:

$$\langle q_0 \lambda \mathbf{q_0} \rangle \to a \langle q_0 X \mathbf{q_0} \rangle$$
$$\langle q_0 \lambda \mathbf{q_1} \rangle \to a \langle q_0 X \mathbf{q_1} \rangle$$

The state in the *third slot* in the variable on the left of each line must be *identical* in the the *third slot* of the *last* variable on the right.

Now consider the rule $a, X \to XX$. This time there will be *two variables* on the right-hand side of the rule. We begin with the following incomplete rule pattern:

$$\langle q_0 X - \rangle \to a \langle q_0 X \square \rangle \langle \square X - \rangle$$

Identical patterns above indicate that the states in those positions must be the same. The single-dash pattern was explained above—it represents where we end up when the after-effects of popping X have vanished from the stack. Part of those after-effects is that two additional X's are pushed and must be subsequently processed. The box patterns must be identical because the state we end up in after processing the first of the two additional X's is where we must continue from to process the second X. We have, therefore, four possible grammar rules for this transition:

$$\langle q_0 X q_0 \rangle \to a \langle q_0 X q_0 \rangle \langle q_0 X q_0 \rangle$$
$$\langle q_0 X q_1 \rangle \to a \langle q_0 X q_0 \rangle \langle q_0 X q_1 \rangle$$
$$\langle \mathsf{q_0 X q_0} \rangle \to \mathsf{a} \langle \mathsf{q_0 X q_1} \rangle \langle \mathsf{q_1 X q_0} \rangle \qquad \text{(invalid)}$$
$$\langle q_0 X q_1 \rangle \to a \langle q_0 X q_1 \rangle \langle q_1 X q_1 \rangle$$

One of those rules, set off in a different typeface, has an invalid variable: $\langle \mathsf{q_1 X q_0} \rangle$. It is not possible to move from state q_1 to state q_0 in this PDA, so we remove this rule from further consideration, leaving only three possible rules for the associated transition.

In like fashion, the transition $b, X \to \lambda$ from q_0 to q_1 yields the following possibilities, one of which we immediately discard because it also reflects an invalid path in the PDA:

$$\langle q_0 X q_0 \rangle \to b\langle q_1 \lambda q_0 \rangle \qquad \text{(invalid)}$$
$$\langle q_0 X q_1 \rangle \to b\langle q_1 \lambda q_1 \rangle$$

Finally, the transition $b, X \to \lambda$ on q_1 gives:

$$\langle q_1 X q_0 \rangle \to b\langle q_1 \lambda q_0 \rangle \qquad \text{(invalid)}$$
$$\langle q_1 X q_1 \rangle \to b\langle q_1 \lambda q_1 \rangle$$

Grouping the surviving rules gives the following candidate grammar:

$$\begin{aligned}
S &\to \langle q_0 \lambda q_0 \rangle \mid \langle q_0 \lambda q_1 \rangle \\
\langle q_0 \lambda q_0 \rangle &\to \lambda \mid a\langle q_0 X q_0 \rangle \\
\langle q_1 \lambda q_1 \rangle &\to \lambda \\
\langle q_0 \lambda q_1 \rangle &\to a\langle q_0 X q_1 \rangle \\
\langle q_0 X q_0 \rangle &\to a\langle q_0 X q_0 \rangle\langle q_0 X q_0 \rangle \\
\langle q_0 X q_1 \rangle &\to a\langle q_0 X q_0 \rangle\langle q_0 X q_1 \rangle \mid a\langle q_0 X q_1 \rangle\langle q_1 X q_1 \rangle \mid b\langle q_1 \lambda q_1 \rangle \\
\langle q_1 X q_1 \rangle &\to b\langle q_1 \lambda q_1 \rangle
\end{aligned}$$

Next, we *simplify* the grammar. We haven't renamed variables yet because $\langle q_0 X q_0 \rangle$ doesn't terminate (no sense in giving it a name), so we will remove any rules containing it, leaving:

$$\begin{aligned}
S &\to \langle q_0 \lambda q_0 \rangle \mid \langle q_0 \lambda q_1 \rangle \\
\langle q_0 \lambda q_0 \rangle &\to \lambda \\
\langle q_1 \lambda q_1 \rangle &\to \lambda \\
\langle q_0 \lambda q_1 \rangle &\to a\langle q_0 X q_1 \rangle \\
\langle q_0 X q_1 \rangle &\to a\langle q_0 X q_1 \rangle\langle q_1 X q_1 \rangle \mid b\langle q_1 \lambda q_1 \rangle \\
\langle q_1 X q_1 \rangle &\to b\langle q_1 \lambda q_1 \rangle
\end{aligned}$$

Next, substitute lambda for $\langle q_0 \lambda q_0 \rangle$ and $\langle q_1 \lambda q_1 \rangle$ everywhere:

$$\begin{aligned}
S &\to \lambda \mid \langle q_0 \lambda q_1 \rangle \\
\langle q_0 \lambda q_1 \rangle &\to a\langle q_0 X q_1 \rangle \\
\langle q_0 X q_1 \rangle &\to a\langle q_0 X q_1 \rangle\langle q_1 X q_1 \rangle \mid b \\
\langle q_1 X q_1 \rangle &\to b
\end{aligned}$$

Next, substitute b for $\langle q_1 X q_1 \rangle$:

$$
\begin{aligned}
S &\rightarrow \lambda \mid \langle q_0 \lambda q_1 \rangle \\
\langle q_0 \lambda q_1 \rangle &\rightarrow a \langle q_0 X q_1 \rangle \\
\langle q_0 X q_1 \rangle &\rightarrow a \langle q_0 X q_1 \rangle b \mid b
\end{aligned}
$$

and then back-substitute $\langle q_0 \lambda q_1 \rangle$:

$$
\begin{aligned}
S &\rightarrow \lambda \mid a \langle q_0 X q_1 \rangle \\
\langle q_0 X q_1 \rangle &\rightarrow a \langle q_0 X q_1 \rangle b \mid b
\end{aligned}
$$

Renaming $\langle q_0 X q_1 \rangle$ as X we obtain the final version of the grammar:

$$
\begin{aligned}
S &\rightarrow \lambda \mid aX \\
X &\rightarrow aXb \mid b
\end{aligned}
$$

It is easy to see that this is equivalent to $S \rightarrow aSb \mid \lambda$, the original grammar for $a^n b^n$.

Example 6–18

We now convert the PDA that recognizes even palindromes in Figure 5–4 to a CFG. First, normalize the PDA:

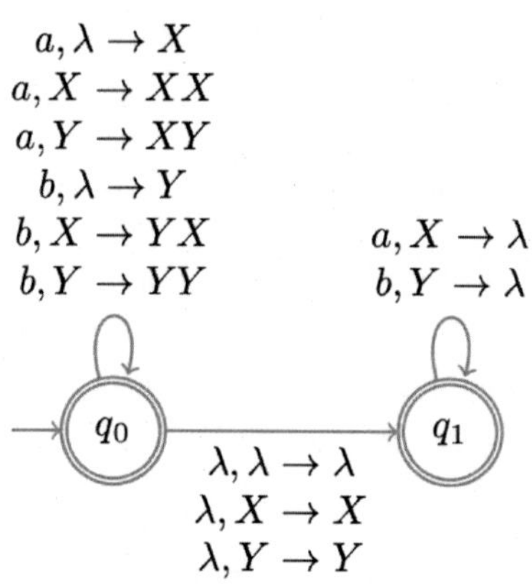

Figure 6–20: Figure 5–4 normalized

The automatic, initial rules are:

$$S \to \langle q_0 \lambda q_0 \rangle \mid \langle q_0 \lambda q_1 \rangle$$
$$\langle q_0 \lambda q_0 \rangle \to \lambda$$
$$\langle q_1 \lambda q_1 \rangle \to \lambda$$

We now extract grammar rules for each transition.

$a, \lambda \to X$ yields the pattern $\langle q_0 \lambda - \rangle \to a \langle q_0 X - \rangle$, giving the rules:

$$\langle q_0 \lambda q_0 \rangle \to a \langle q_0 X q_0 \rangle$$
$$\langle q_0 \lambda q_1 \rangle \to a \langle q_0 X q_1 \rangle$$

$a, X \to XX$ has the pattern $\langle q_0 X - \rangle \to a \langle q_0 X \Box \rangle \langle \Box X - \rangle$, giving:

$$\langle q_0 X q_0 \rangle \to a \langle q_0 X q_0 \rangle \langle q_0 X q_0 \rangle$$
$$\langle q_0 X q_0 \rangle \to a \langle q_0 X q_1 \rangle \langle q_1 X q_0 \rangle \qquad \text{(invalid)}$$
$$\langle q_0 X q_1 \rangle \to a \langle q_0 X q_0 \rangle \langle q_0 X q_1 \rangle$$
$$\langle q_0 X q_1 \rangle \to a \langle q_0 X q_1 \rangle \langle q_1 X q_1 \rangle$$

$a, Y \to XY$ has the pattern $\langle q_0 Y - \rangle \to a \langle q_0 X \Box \rangle \langle \Box Y - \rangle$, giving:

$$\langle q_0 Y q_0 \rangle \to a \langle q_0 X q_0 \rangle \langle q_0 Y q_0 \rangle$$
$$\langle q_0 Y q_0 \rangle \to a \langle q_0 X q_1 \rangle \langle q_1 Y q_0 \rangle \qquad \text{(invalid)}$$
$$\langle q_0 Y q_1 \rangle \to a \langle q_0 X q_0 \rangle \langle q_0 Y q_1 \rangle$$
$$\langle q_0 Y q_1 \rangle \to a \langle q_0 X q_1 \rangle \langle q_1 Y q_1 \rangle$$

$b, \lambda \to Y$ has pattern $\langle q_0 \lambda - \rangle \to b \langle q_0 Y - \rangle$, giving:

$$\langle q_0 \lambda q_0 \rangle \to b \langle q_0 Y q_0 \rangle$$
$$\langle q_0 \lambda q_1 \rangle \to b \langle q_0 Y q_1 \rangle$$

$b, X \to YX$ has pattern $\langle q_0 X - \rangle \to b \langle q_0 Y \Box \rangle \langle \Box X - \rangle$, giving:

$$\langle q_0 X q_0\rangle \to b\langle q_0 Y q_0\rangle\langle q_0 X q_0\rangle$$
$$\langle q_0 X q_0\rangle \to b\langle q_0 Y q_1\rangle\langle q_1 X q_0\rangle \qquad \text{(invalid)}$$
$$\langle q_0 X q_1\rangle \to b\langle q_0 Y q_0\rangle\langle q_0 X q_1\rangle$$
$$\langle q_0 X q_1\rangle \to b\langle q_0 Y q_1\rangle\langle q_1 X q_1\rangle$$

$b, Y \to YY$ has pattern $\langle q_0 Y-\rangle \to b\langle q_0 Y\Box\rangle\langle\Box Y-\rangle$, giving:

$$\langle q_0 X q_0\rangle \to b\langle q_0 Y q_0\rangle\langle q_0 Y q_0\rangle$$
$$\langle q_0 Y q_0\rangle \to b\langle q_0 Y q_1\rangle\langle q_1 Y q_0\rangle \qquad \text{(invalid)}$$
$$\langle q_0 Y q_1\rangle \to b\langle q_0 Y q_0\rangle\langle q_0 Y q_1\rangle$$
$$\langle q_0 Y q_1\rangle \to b\langle q_0 Y q_1\rangle\langle q_1 Y q_1\rangle$$

$\lambda, \lambda \to \lambda$ has pattern $\langle q_0 \lambda-\rangle \to \langle q_1 \lambda-\rangle$, giving:

$$\langle q_0 \lambda q_0\rangle \to \langle q_1 \lambda q_0\rangle \qquad \text{(invalid)}$$
$$\langle q_0 \lambda q_1\rangle \to \langle q_1 \lambda q_1\rangle$$

$\lambda, X \to X$ has pattern $\langle q_0 X-\rangle \to \langle q_1 X-\rangle$, giving:

$$\langle q_0 X q_0\rangle \to \langle q_1 X q_0\rangle \qquad \text{(invalid)}$$
$$\langle q_0 X q_1\rangle \to \langle q_1 X q_1\rangle$$

$\lambda, Y \to X$ has pattern $\langle q_0 Y-\rangle \to \langle q_1 X-\rangle$, giving:

$$\langle q_0 Y q_0\rangle \to \langle q_1 X q_0\rangle \qquad \text{(invalid)}$$
$$\langle q_0 Y q_1\rangle \to \langle q_1 X q_1\rangle$$

$a, X \to \lambda$ has pattern $\langle q_1 X-\rangle \to \langle q_1 \lambda-\rangle$, giving:

$$\langle q_0 X q_0\rangle \to \langle q_1 \lambda q_0\rangle \qquad \text{(invalid)}$$
$$\langle q_0 X q_1\rangle \to \langle q_1 \lambda q_1\rangle$$

Finally, $b, Y \to \lambda$ has pattern $\langle q_1 Y-\rangle \to \langle q_1 \lambda-\rangle$, giving:

$$\langle q_0 Y q_0 \rangle \rightarrow \langle q_1 \lambda q_0 \rangle \qquad \text{(invalid)}$$

$$\langle q_0 Y q_1 \rangle \rightarrow \langle q_1 \lambda q_1 \rangle$$

Now gather the valid rules together:

$$\begin{aligned}
S &\rightarrow \langle q_0 \lambda q_0 \rangle \mid \langle q_0 \lambda q_1 \rangle \\
\langle q_0 \lambda q_0 \rangle &\rightarrow \lambda \mid a\langle q_0 X q_0 \rangle \mid b\langle q_0 Y q_0 \rangle \\
\langle q_1 \lambda q_1 \rangle &\rightarrow \lambda \\
\langle q_0 \lambda q_1 \rangle &\rightarrow a\langle q_0 X q_1 \rangle \mid b\langle q_0 Y q_1 \rangle \mid \langle q_1 \lambda q_1 \\
\langle q_0 X q_0 \rangle &\rightarrow a\langle q_0 X q_0 \rangle\langle q_0 X q_0 \rangle \mid b\langle q_0 Y q_0 \rangle\langle q_0 X q_0 \rangle \\
\langle q_0 X q_1 \rangle &\rightarrow a\langle q_0 X q_1 \rangle\langle q_1 X q_1 \rangle \mid a\langle q_0 X q_0 \rangle\langle q_0 X q_1 \rangle \mid b\langle q_0 Y q_0 \rangle\langle q_0 X q_1 \rangle \\
&\quad \mid b\langle q_0 Y q_1 \rangle\langle q_1 X q_1 \mid \langle q_1 X q_1 \rangle \\
\langle q_0 Y q_0 \rangle &\rightarrow a\langle q_0 X q_0 \rangle\langle q_0 Y q_0 \rangle \mid b\langle q_0 Y q_0 \rangle\langle q_0 Y q_0 \rangle \\
\langle q_0 Y q_1 \rangle &\rightarrow a\langle q_0 Y q_1 \rangle\langle q_0 Y q_1 \rangle \mid a\langle q_0 X q_1 \rangle\langle q_1 Y q_1 \rangle \mid b\langle q_0 Y q_0 \rangle\langle q_0 X q_1 \rangle \\
&\quad \mid b\langle q_0 Y q_1 \rangle\langle q_1 Y q_1 \mid \langle q_1 Y q_1 \rangle \\
\langle q_1 X q_1 \rangle &\rightarrow a\langle q_1 \lambda q_1 \rangle \\
\langle q_1 Y q_1 \rangle &\rightarrow b\langle q_1 \lambda q_1 \rangle
\end{aligned}$$

Neither $\langle q_0 X q_0 \rangle$ nor $\langle q_0 Y q_0 \rangle$ terminate, so remove all rules containing them. This leaves $\langle q_0 \lambda q_0 \rangle$ reduced to lambda, which can now be back-substituted. Also, the fact that $\langle q_1 \lambda q_1 \rangle$ is empty implies that $\langle q_1 X q_1 \rangle \rightarrow a$ and $\langle q_1 Y q_1 \rangle \rightarrow b$. After making all these changes the grammar becomes:

$$\begin{aligned}
S &\rightarrow \lambda \mid \langle q_0 \lambda q_1 \rangle \\
\langle q_0 \lambda q_1 \rangle &\rightarrow a\langle q_0 X q_1 \rangle \mid b\langle q_0 Y q_1 \rangle \mid \lambda \\
\langle q_0 X q_1 \rangle &\rightarrow a\langle q_0 X q_1 \rangle a \mid b\langle q_0 Y q_1 \rangle a \mid a \\
\langle q_0 Y q_1 \rangle &\rightarrow a\langle q_0 X q_1 \rangle b \mid b\langle q_0 Y q_1 \rangle b \mid b
\end{aligned}$$

Renaming $S \leftrightarrow \langle q_0 \lambda q_1 \rangle$, $X \leftrightarrow \langle q_0 X q_1 \rangle$, and $Y \leftrightarrow \langle q_0 Y q_1 \rangle$, we get the final, simplified grammar:

$$\begin{aligned}
S &\rightarrow \lambda \mid aX \mid bY \\
X &\rightarrow aXa \mid bYa \mid a \\
Y &\rightarrow aXb \mid bYb \mid b
\end{aligned}$$

As a sanity check we derive $abbbba$:

$$S \Rightarrow aX \Rightarrow abYa \Rightarrow abbYba \Rightarrow abbbba$$

Conversion from PDA to CFG (General Case)

1. Use a normalized PDA.
2. Form rules $S \to \langle s\lambda f\rangle$ for start state s and *every accepting state* f.
3. Form rules $\langle q\lambda q\rangle \to \lambda$ for *every state* q.
4. For *every PDA transition*, $c, V \to s$, from state p to state q, where s is the string of pushed stack symbols, $V_1V_2\cdots V_k$, form the following rules:

$$\langle pVr_k\rangle \to c\langle qV_1r_1\rangle\langle r_1V_2r_2\rangle\langle r_2V_3r_3\rangle\cdots\langle r_{k-2}V_{k-1}r_{k-1}\rangle\langle r_{k-1}V_kr_k\rangle$$

for all combinations of states r_i in the PDA. You may remove any rule with variables $\langle r_jV_{j+1}r_{j+1}\rangle$ where movement from state r_j to state r_{j+1} is not possible. Simplify the grammar as desired.

Key Terms

lambda-pop transitions • after-effects of a stack variable • normalized PDA

Exercises

1. Convert the following CFG to a PDA.

$$\begin{aligned} S &\to \lambda \mid aB \mid bA \\ A &\to aS \mid bAA \\ B &\to bS \mid aBB \end{aligned}$$

2. Convert the following single-state PDA to a CFG.

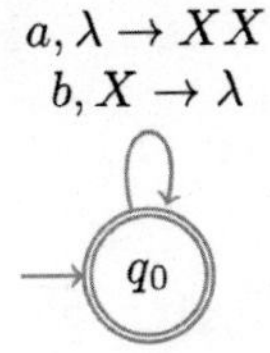

3. Convert the following two-state PDA to a CFG.

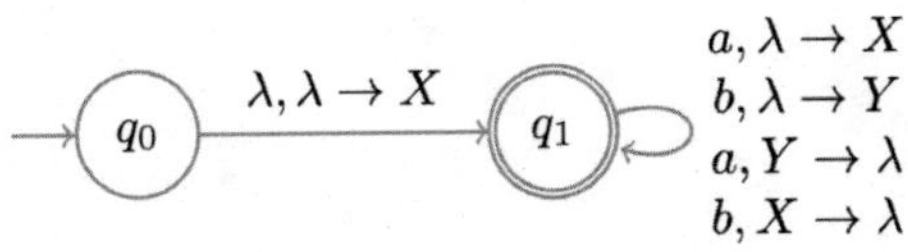

Chapter Summary

Context-free grammars generate a larger class of languages than finite automata. Regular languages are the subset of context-free languages that do not use the stack. They are useful in natural language processing and in parsing programming languages during compilation. Operator precedence and associativity is evident in the derivation tree of an expression, and is also reflected by positioning of rules and recursion of variables in the CFG. A grammar that allows multiple, distinct derivation trees (or left most derivations) for the *same string* is an ambiguous grammar. We have shown that any CFG has an associated PDA that accepts the same language, and vice-versa; therefore, CFGs and PDAs both describe the class of context-free languages.

7. Properties of Context-Free Languages

You've got to develop relationships. You can't do things just in a formal context.

– Kenneth Chenault

"Beach Properties", by thekirbster Licensed under CC-BY 2.0

Where Are We?

Language Class	Recognized By	Generated By
Regular	Finite Automata	Regular Grammars Regular Expressions
Context-Free	Pushdown Automata	Context-Free Grammars
Recursively Enumerable	Turing Machines	Unrestricted Grammars

Chapter Objectives

- *Convert Context-Free Grammars to Chomsky Normal Form*
- *Determine which operations preserve the "context-free-ness" of a language*
- *Learn algorithms for answering fundamental questions about Context-Free Languages*
- *Show that some languages are not context-free*

As with Chapter 4 for regular languages, this chapter takes a deeper look at the closure and other properties of context-free languages, develops algorithms for making decisions about any context-free language, and finally, introduces a formal language that is not context-free, eventually leading to the third and final class of formal languages. We begin with a procedure for converting any CFG into a special form, **Chomsky Normal Form** (CNF), which will facilitate deriving certain results about Context-free Languages.

7.1 Chomsky Normal Form

In preparation for Chomsky Normal Form, we need to make two modifications to any CFG under consideration:

1. Remove *lambda* from all rules without changing the language generated (except we may lose the empty string, which is okay)
2. Remove *unit productions*, such as $X \to Y$, while maintaining their effects

There are practical reasons for removing lambda from a CFG. To process a string efficiently to see whether it is generated by a CFG, it is better if the intermediate sentential forms of the working derivation not decrease in size. If lambda is a valid replacement for a variable, then the working derivation string can shrink, so it is possible that a working derivation can grow and shrink and grow and shrink repeatedly, which is inefficient. A grammar can maintain the effect of the empty string in its rules without explicitly using lambda, so that the sentential forms don't shrink. Such a grammar is called **non-contracting**. The only downside is that the modified grammar will no longer generate the empty string, but that is a special case not worth worrying about most of the time.

Removing Lambda

To remove lambda while maintaining its contribution in forming non-empty strings, we first identify which variables can generate the empty string, whether directly or indirectly. Such variables are called **nullable variables**, and, once identified, we incorporate their *nullable effects* into the rules of the grammar, removing any explicit mention of lambda. The language recognized by the modified grammar will be $L - \{\lambda\}$, where L represents the language recognized by the original grammar.

To illustrate, we remove lambda from the following grammar.

$$S \to a \mid Xb \mid aYa$$
$$X \to Y \mid \lambda$$
$$Y \to b \mid X$$

X is the only variable that goes to lambda directly, but Y is also nullable because of the production $Y \to X$. To substitute the effects of both X and Y being nullable, we change the grammar as follows:

$$\begin{aligned} S &\to a \mid Xb \mid aYa \mid b \mid aa \\ X &\to Y \mid \lambda \\ Y &\to b \mid X \end{aligned}$$

In the first line, we added the rule b from substituting lambda for X in the rule $S \to Xb$. Likewise, we obtain the rule $S \to aa$ from $S \to aYa$ since Y is nullable. The goal is to eliminate lambda completely, so we don't bother to substitute lambda for X or Y in the last two lines. Having preserved the effects of the nullable variables, we now eliminate lambda:

$$\begin{aligned} S &\to a \mid Xb \mid aYa \mid b \mid aa \\ X &\to Y \\ Y &\to b \mid X \end{aligned}$$

Since the grammar did not generate the empty string to begin with, the latter grammar generates the same language as the original. If desired, we can further simplify this grammar by eliminating X, since it is equivalent to Y:

$$\begin{aligned} S &\to a \mid Yb \mid aYa \mid b \mid aa \\ Y &\to b \mid Y \end{aligned}$$

It is useless to have a "self-rule" like $Y \to Y$, so we drop that as well:

$$\begin{aligned} S &\to a \mid Yb \mid aYa \mid b \mid aa \\ Y &\to b \end{aligned}$$

Finally, we back-substitute for $Y \to b$ to obtain:

$$S \to a \mid bb \mid aba \mid b \mid aa$$

It wasn't evident upon inspection that the original grammar generated such a simple, finite language.

While it was straightforward to discover that Y was indirectly nullable in the grammar above, more complex grammars call for an *algorithm* to discover all nullable variables. The following recursive definition facilitates discovering such an algorithm.

Definition 7.1

- A variable, X, is **nullable** if there is rule of the form $X \to \lambda$, or
- A variable, Y, is **nullable** if there is rule of the form $Y \to X_1X_2 \cdots X_k$ where all of the variables X_i are nullable.

To illustrate the algorithm on the grammar above, we initialize our set of nullable variables, $\mathcal{N}$, to be the set $\{X\}$, because of the rule $X \to \lambda$. We now use part 2 of the definition to find rules that consist solely of a right-hand of X's, and discover $Y \to X$, so we add Y to our set, giving $\mathcal{N} = \{X, Y\}$. Next, we look for rules that have only a combination of X's and/or Y's as right-hand sides. Seeing none, we conclude that only X and Y are nullable.

A more visual technique to obtain the same result is to modify a working copy of the grammar by substituting lambda for all directly nullable variables in the grammar, and then repeat on the modified grammar until we obtain no more new nullables. For this case we start with the original grammar and the knowledge that X is nullable from the rule $X \to \lambda$, and substitute lambda for X everywhere:

$$\begin{aligned} S &\to a \mid b \mid aYa \\ X &\to Y \mid \lambda \\ Y &\to b \mid \lambda \end{aligned}$$

This reveals that Y is also nullable by the rule $Y \to \lambda$, so we repeat the lambda-substitution for Y:

$$\begin{aligned} S &\to a \mid b \mid aa \\ X &\to \lambda \\ Y &\to b \mid \lambda \end{aligned}$$

No new directly nullable relationships appear, so we end with $\mathcal{N} = \{X, Y\}$ as before. *We are not creating a new grammar here*—we are merely using a copy of the grammar as a temporary workspace for the algorithm. What we ended up with above is not equivalent to the original grammar.

There is also a useful trick to assist in *substituting* the nullable effects of a variable throughout a grammar. For each nullable variable, V, we substitute the expression $(V \mid \lambda)$ for V everywhere it appears in the right-hand side of a rule (except when V appears all by itself), and then distribute the expression per the algebra for regular expressions. For the current example, we perform the suitable substitutions for nullables X and Y below in the original grammar.

$$\begin{aligned} S &\to a \mid (X \mid \lambda)b \mid a(Y \mid \lambda)a \\ X &\to Y \mid \lambda \\ Y &\to b \mid X \end{aligned}$$

Now expand the expressions in the first line:

$$\begin{aligned} S &\to a \mid Xb \mid \mathbf{b} \mid aYa \mid \mathbf{aa} \\ X &\to Y \mid \lambda \\ Y &\to b \mid X \end{aligned}$$

Example 7–1

Remove lambda, while maintaining its effects, from the grammar below.

$$\begin{aligned} S &\to XYZ \\ X &\to aXb \mid Y \\ Y &\to bY \mid \lambda \\ Z &\to ccZ \mid Y \end{aligned}$$

Discover the nullable variables by first initializing $\mathcal{N} = \{Y\}$, and substituting lambda for Y in all rules:

$$\begin{aligned} S &\to XZ \\ X &\to aXb \mid \lambda \\ Y &\to b \mid \lambda \\ Z &\to ccZ \mid \lambda \end{aligned}$$

It is evident that X and Z are nullable as well, so $\mathcal{N} = \{X, Y, Z\}$, and substitute lambda everywhere for these two variables:

$$\begin{aligned} S &\to \lambda \\ X &\to ab \mid \lambda \\ Y &\to b \mid \lambda \\ Z &\to cc \mid \lambda \end{aligned}$$

All variables are nullable, so $\mathcal{N} = \{S, X, Y, Z\}$ (although S doesn't appear in any right-hand-side rule). Now substitute the expression $(V \mid \lambda)$ for all variables $V \in \mathcal{N}$ wherever they appear embedded inside rules of the original grammar:

$$\begin{aligned} S &\to (X \mid \lambda)(Y \mid \lambda) \\ X &\to a(X \mid \lambda)b \mid Y \\ Y &\to b(Y \mid \lambda) \mid \lambda \\ Z &\to cc(Z \mid \lambda) \mid Y \end{aligned}$$

Distributing these through and eliminating lambda yields the following grammar for $L - \{\lambda\}$:

$$\begin{aligned} S &\to XYZ \mid XY \mid XZ \mid YZ \mid X \mid Y \mid Z \\ X &\to aXb \mid ab \mid Y \\ Y &\to bY \mid b \\ Z &\to ccZ \mid cc \mid Y \end{aligned}$$

Since S does not appear in any right-hand-side rule, we can add lambda back in as a choice for S and preserve the empty string from the original language, if desired.

Removing Unit Productions

Unit productions are inefficient in that using one in a working derivation does not grow the intermediate sentential form. It would be better to have derivations that monotonically increase in length. Instead of using the rule $X \to Y$, we prefer to "skip the middleman" and substitute directly what Y yields.

To illustrate, look at the final version of the non-contracting grammar in Example 7–1 above, which contains the following unit productions:

$$\begin{aligned} S &\to X \mid Y \mid Z \\ X &\to Y \\ Z &\to Y \end{aligned}$$

The following **unit production graph** helps visualize how unit productions affect a grammar.

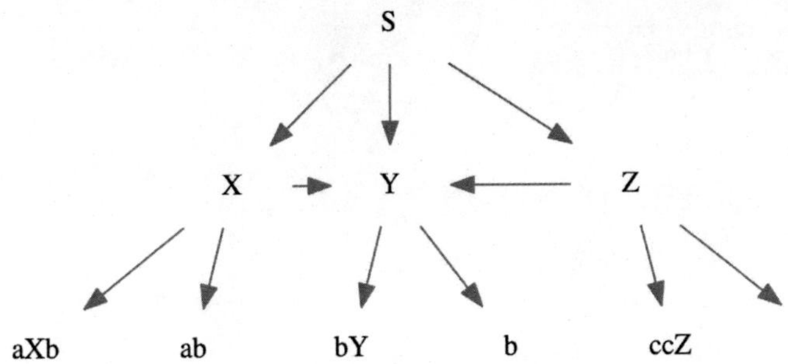

Figure 7–1: Unit Production Graph for Example 7–1

Wherever X appears as a unit production, it can be replaced not only with its two right-hand sides, but with Y's two non-unit strings as well. The same goes for Z. We call such replacement paths **unit chains**. Here are all the unit chains from the graph above:

$$
\begin{aligned}
S \to X \to Y \\
S \to Y \\
S \to Z \to Y \\
X \to Y \\
Z \to Y
\end{aligned}
$$

From here we can deduce all the variables reachable by unit productions. For example, if S appears alone on the right-hand side of a rule, we can replace it with the *non-unit* effects of S, X, Y, and Z, because X, Y and Z are reachable from S by unit productions. These variables constitute the **unit chain set** of S. The chain sets for this grammar are:

$$
\begin{aligned}
\text{chain}(S) &= \{S, X, Y, Z\} \\
\text{chain}(X) &= \{X, Y\} \\
\text{chain}(Y) &= \{Y\} \\
\text{chain}(Z) &= \{Z, Y\}
\end{aligned}
$$

So we replace X as a unit production with $aXb \mid ab \mid bY \mid b$, which are the non-unit replacements for the variables in $\text{chain}(X) = \{X, Y\}$. Likewise, when Z appears alone on the right-hand side of a rule, we replace it with the non-unit replacements of Z and Y. Replacing all the unit productions with the non-unit effects of their respective chain sets gives us:

$$
\begin{aligned}
S &\rightarrow XYZ \mid XY \mid XZ \mid YZ \mid \mathbf{aXb} \mid \mathbf{ab} \mid \mathbf{bY} \mid \mathbf{b} \mid \mathbf{ccZ} \mid \mathbf{cc} \\
X &\rightarrow aXb \mid ab \mid \mathbf{bY} \mid \mathbf{b} \\
Y &\rightarrow bY \mid b \\
Z &\rightarrow ccZ \mid cc \mid bY \mid b
\end{aligned}
$$

Example 7–2

The grammar below also has chains of unit productions.

$$
\begin{aligned}
S &\rightarrow ACA \mid CA \mid AA \mid AC \mid A \mid C \\
A &\rightarrow aAa \mid aa \mid B \mid C \\
B &\rightarrow bB \mid C \\
C &\rightarrow cC \mid c
\end{aligned}
$$

For clarity, we omit the non-unit strings to discover the chains in the unit production graph below:

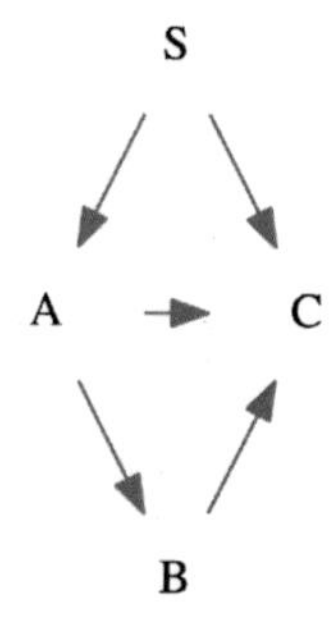

Figure 7–2: Unit production graph with non-unit rules omitted

This gives the following chain sets:

$$
\begin{aligned}
\text{chain}(S) &= \{S, A, B, C\} \\
\text{chain}(A) &= \{A, B, C\} \\
\text{chain}(B) &= \{B, C\} \\
\text{chain}(C) &= \{C\}
\end{aligned}
$$

Wherever these variables appear alone as a unit on the right-hand side of a rule, we replace them with all non-unit rules for every variable in their chain set. In this case, it is convenient to substitute C's non-unit effects into B's rule first, giving $B \rightarrow bB \mid cC \mid c$. Then we substitute the new B into A's right-hand side, and then the new A into S's rule, resulting in the following grammar, dropping duplicate substitutions as needed:

$$S \to ACA \mid CA \mid AA \mid AC \mid \mathbf{aAa} \mid \mathbf{aa} \mid \mathbf{bB} \mid \mathbf{cC} \mid \mathbf{c}$$
$$A \to aAa \mid aa \mid \mathbf{bB} \mid \mathbf{cC} \mid \mathbf{c}$$
$$B \to bB \mid \mathbf{cC} \mid \mathbf{c}$$
$$C \to cC \mid c$$

It is important to remove null rules *before* unit rules. To see why, look at the grammar below. It has no unit productions, so we will eliminate lambda.

$$S \to aA$$
$$A \to BB$$
$$B \to aBb \mid \lambda$$

Both A and B are nullable, so we substitute accordingly:

$$S \to a(A \mid \lambda) \to aA \mid a$$
$$A \to BB \to (B \mid \lambda)(B \mid \lambda) \to BB \mid \mathbf{B}$$
$$B \to a(B \mid \lambda)b \to aBb \mid ab$$

Removing lambda in this case introduced a new unit production, $A \to B$. Always remove lambda first.

Remove lambda before removing unit productions.

Chomsky Normal Form Rules

When a grammar does not use lambda in any rule and has no unit productions, it is ready to convert to Chomsky Normal Form (CNF). CNF allows rules of only two types:

1. $X \to c$, where $X \in V$, and $c \in \Sigma$
2. $X \to YZ$, where $X, Y, Z \in V$

In other words, right-hand sides may only have a single (non-empty) terminal symbol or exactly two variables. If applicable, we can retain the empty string with the rule $S \to \lambda$, but only if S does not appear on the right-hand side of any rule.

Sometimes we can remove S from a replacement rule. Recall the grammar for $a^n b^n$, $S \to aSb \mid \lambda$. We can remove S from the right-hand side as follows:

$$
\begin{aligned}
S &\to aXb \mid ab \mid \lambda \\
X &\to aXb \mid ab
\end{aligned}
$$

If we desire a string other than λ or ab, we use the variable X. To convert this to Chomsky Normal Form, we first introduce new arbitrarily named variables, A and B:

$$
\begin{aligned}
S &\to AXB \mid AB \mid \lambda \\
X &\to AXB \mid AB \\
A &\to a \\
B &\to b
\end{aligned}
$$

The only right-hand side not in the proper form is AXB, so we introduce another variable, W, with the rule $W \to AX$ to obtain the Chomsky Normal Form for the original grammar:

$$
\begin{aligned}
S &\to WB \mid AB \mid \lambda \\
X &\to WB \mid AB \\
A &\to a \\
B &\to b \\
W &\to AX
\end{aligned}
$$

The choice of variable pairs to replace is also arbitrary, so we could also have defined $W \to XB$ to obtain an equivalent CNF grammar:

$$
\begin{aligned}
S &\to AW \mid AB \mid \lambda \\
X &\to AW \mid AB \\
A &\to a \\
B &\to b \\
W &\to XB
\end{aligned}
$$

After removing lambda (except for $S \to \lambda$, as explained above) and unit productions, there are two types of transformations needed to form CNF-style rules:

1. Wherever a terminal appears in a rule of length greater than 1, replace each terminal with a unique non-terminal. For example, for the right-hand side aAa, we could replace it with XAX, say, and add the rule $X \rightarrow a$ to the grammar (where X is a *new* variable).
2. After completing step 1, split right-hand sides consisting of more than two variables into two substrings, and recursively introduce new variables that map to each substring until every rule has only two variables.

These steps must be applied in order. We illustrate by applying a Type-1 transformation to the final form of the grammar in Example 7–2, adding new variables X, Y, and Z:

$$
\begin{aligned}
&S \rightarrow ACA \mid CA \mid AA \mid AC \mid XAX \mid XX \mid YB \mid ZC \\
&A \rightarrow XAX \mid XX \mid YB \mid ZC \\
&B \rightarrow YB \mid ZC \\
&C \rightarrow ZC \mid c \\
&X \rightarrow a \\
&Y \rightarrow b \\
&Z \rightarrow c
\end{aligned}
$$

Now we need to split the three rules with length greater than 2, namely:

$$
\begin{aligned}
&S \rightarrow ACA \mid XAX \\
&A \rightarrow XAX
\end{aligned}
$$

We arbitrarily split each right-hand side into two parts, assigning a new variable for the part of length two. For example, we can split ACA into AC and A, and introduce the rule $W \rightarrow AC$, rewriting the original rule as $S \rightarrow WA$. We can also replace the right-hand side XAX with VX where $V \rightarrow XA$. With these changes, we render the grammar in CNF as follows.

$$
\begin{aligned}
S &\rightarrow WA \mid CA \mid AA \mid AC \mid VX \mid XX \mid YB \mid ZC \\
A &\rightarrow VX \mid XX \mid YB \mid ZC \\
B &\rightarrow YB \mid ZC \\
C &\rightarrow ZC \mid c \\
X &\rightarrow a \\
Y &\rightarrow b \\
Z &\rightarrow c \\
W &\rightarrow AC \\
V &\rightarrow XA
\end{aligned}
$$

Example 7–3

Convert the following grammar to CNF.

$$
\begin{aligned}
S &\rightarrow bA \mid aB \\
A &\rightarrow bAA \mid aS \mid a \\
B &\rightarrow aBB \mid bS \mid b
\end{aligned}
$$

There are no lambdas or unit productions, so we begin by adding rules for the terminals:

$$
\begin{aligned}
S &\rightarrow YA \mid XB \\
A &\rightarrow YAA \mid XS \mid a \\
B &\rightarrow XBB \mid YS \mid b \\
X &\rightarrow a \\
Y &\rightarrow b
\end{aligned}
$$

Finally, we split the rules greater than 2 in length to obtain CNF:

$$\begin{aligned}
S &\rightarrow YA \mid XB \\
A &\rightarrow ZA \mid XS \mid a \\
B &\rightarrow WB \mid YS \mid b \\
X &\rightarrow a \\
Y &\rightarrow b \\
Z &\rightarrow YA \\
W &\rightarrow XB
\end{aligned}$$

Example 7–4

The following grammar needs lambdas and unit productions removed before converting it to CNF.

$$\begin{aligned}
S &\rightarrow abAB \\
A &\rightarrow aAB \mid \lambda \\
B &\rightarrow BAa \mid A \mid \lambda
\end{aligned}$$

Both A and B are nullable, so distribute their nullable effects throughout the grammar and drop lambda:

$$\begin{aligned}
S &\rightarrow ab(A \mid \lambda)(B \mid \lambda) \rightarrow abAB \mid abA \mid abB \mid ab \\
A &\rightarrow b(A \mid \lambda)(B \mid \lambda) \rightarrow bAB \mid bA \mid bB \mid b \\
B &\rightarrow (B \mid \lambda)(A \mid \lambda)a \mid A \rightarrow BAa \mid Ba \mid Aa \mid a \mid A
\end{aligned}$$

The only unit production is $B \rightarrow A$, so substitute A's non-unit replacements in its place on B's line:

$$\begin{aligned}
S &\rightarrow abAB \mid abA \mid abB \mid ab \\
A &\rightarrow bAB \mid bA \mid bB \mid b \\
B &\rightarrow BAa \mid Ba \mid Aa \mid a \mid bAB \mid bA \mid bB \mid b
\end{aligned}$$

Now add rules for the terminals:

$$
\begin{aligned}
S &\to XYAB \mid XYA \mid XYB \mid XY \\
A &\to YAB \mid YA \mid YB \mid b \\
B &\to BAX \mid BX \mid AX \mid a \mid YAB \mid YA \mid YB \mid b \\
X &\to a \\
Y &\to b
\end{aligned}
$$

For the final step, split the long rules:

$$
\begin{aligned}
S &\to ZW \mid ZA \mid ZB \mid XY \\
A &\to YW \mid YA \mid YB \mid b \\
B &\to TX \mid BX \mid AX \mid a \mid YW \mid YA \mid YB \mid b \\
X &\to a \\
Y &\to b \\
Z &\to XY \\
W &\to AB \\
T &\to BA
\end{aligned}
$$

The constrained nature of Chomsky Normal Form will make it easier to obtain important results in Sections 7.3 and 7.4.

Key Terms

nullable variable • unit production • unit chain • chain set • Chomsky normal form

Exercises

1. Find a context-free grammar that doesn't use lambda that accepts the same language as the following CFG (except for the empty string).

$$
\begin{aligned}
S &\to BSA \mid A \\
A &\to aA \mid \lambda \\
B &\to Bba \mid \lambda
\end{aligned}
$$

2. Find a grammar that accepts the same language as the following CFG, but without unit productions.

$$
\begin{aligned}
S &\to AS \mid C \\
A &\to aA \mid bB \mid C \\
B &\to bB \mid b \\
C &\to cC \mid B
\end{aligned}
$$

3. Convert the following grammar to Chomsky Normal Form (CNF).

$$
\begin{aligned}
S &\to aA \mid BAa \\
A &\to AA \mid a \\
B &\to AbB \mid b
\end{aligned}
$$

4. Convert the following grammar to CNF.

$$
\begin{aligned}
S &\to XY \\
X &\to aXb \mid \lambda \\
Y &\to aY \mid \lambda
\end{aligned}
$$

7.2 Closure Properties

Since there is a difference between deterministic and non-deterministic context-free languages, there are differences in their closure properties. We first develop the closure properties for non-deterministic context-free languages, using CFG-based arguments. "Context-free-ness" is preserved under the following operations:

1. Union
2. Concatenation

3. Kleene star
4. Reversal
5. Regular intersection (intersection with a regular language), and regular union

Suppose $S1$ and $S2$ are the respective start variables for context-free languages L_1 and L_2. We can form a CFG for $L_1 \cup L_2$ as follows.

$$S \to S_1 \mid S_2$$
$$S_1 \to \cdots$$
$$S_2 \to \cdots$$

We introduced a new start variable, S, which recognizes the union of the two languages by providing a choice the two original grammars. The rules for both grammars are part of the combined grammar, with variables renamed as needed to avoid name clashes.

A CFG for the concatenation L_1L_2 is equally straightforward:

$$S \to S_1S_2$$
$$S_1 \to \cdots$$
$$S_2 \to \cdots$$

To form a CFG for the Kleene star of L_1, we form a new grammar that generates the empty string and uses recursion for repetition:

$$S \to S_1S \mid \lambda$$
$$S_1 \to \cdots$$

For reversal, we need only reverse all the right-hand sides in a grammar. Consider, for example the grammar $S \to aSb \mid \lambda$ for the language a^nb^n. Reversing the right-hand sides gives the grammar $S \to bSa \mid \lambda$ which generates the language b^na^n.

Intuitively it makes sense that CFLs would not be closed under intersection in general, since it is unlikely that the stack operations of their respective PDAs would be compatible. The easiest way to show that CFLs are not closed under intersection is by means of a counterexample, where the intersection of two CFLs is not context-free. We will prove in Section 7.4 that the language $a^nb^nc^n$ is not context-free. Accepting that fact for now, note that for CFLs $L_1 = a^nb^nc^m$ and $L_2 = a^mb^nc^n$, their intersection, $L_1 \cap L_2 = a^nb^nc^n$, is not context-free, so in general, CFLs are *not* closed under *intersection.*

CFLs are closed under *regular intersection,* however. Since regular languages do not use the stack, we can construct the intersection of a CFL with a regular language by using the product-table approach from Section 4.1. We will show that the intersection of a^nb^n with the language of strings with an even number of a's and b's is context-free. We begin with transition tables for the two languages (review Table 5-4 for a sample transition table for a CFL). See the machines and tables below.

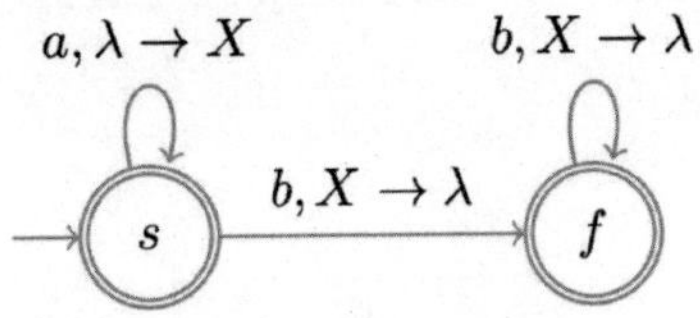

Figure 7–3: PDA for $a^n b^n$

Table 7–1: Transition table for $a^n b^n$

	a, λ	b, X
$+s$	s/X	f/λ
$+f$	—	f/λ

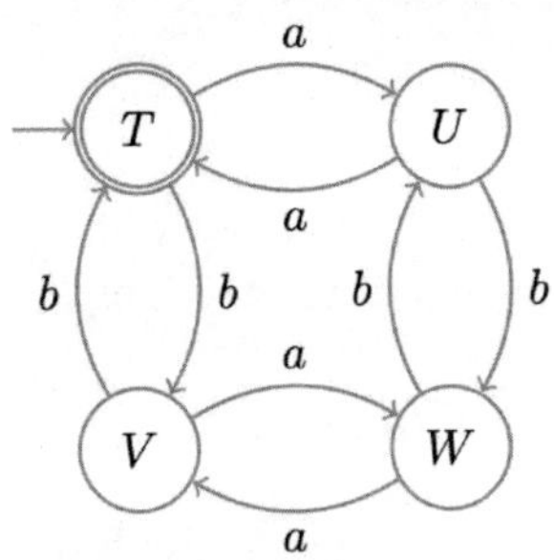

Figure 7–4: DFA for EVEN-EVEN

Table 7–2: Transition table for EVEN-EVEN

	a	b
$+T$	U	V
U	T	W
V	W	T
W	V	U

We now construct the product table, where the column headings are the same as in the transition table for the PDA. As usual, we begin with the composite start state and see where it leads.

Table 7–3: Start of product table

	a, λ	b, X

Table 7–3: Start of product table

	a, λ	b, X
$\{s, T\}$	$\{s/X, U\}$	$\{f/\lambda, V\}$

We see that we have reached the new composite states, $\{s, U\}$ and $\{f, V\}$:

Table 7–4: Next steps in the product table

	a, λ	b, X
$\{s, T\}$	$\{s/X, U\}$	$\{f/\lambda, V\}$
$\{s, U\}$	$\{s/X, T\}$	$\{f/\lambda, W\}$
$\{f/V\}$	$\{-, W\}$	$\{f/\lambda, T\}$

Note that there is no transition in the PDA from state f with input a, λ. This represents a *jail state* in the PDA, which we customarily omit in the result. The completed product table appears below.

Table 7–5: The finished product table

	a, λ	b, X
$\{s, T\}$	$\{s/X, U\}$	$\{f/\lambda, V\}$
$\{s, U\}$	$\{s/X, T\}$	$\{f/\lambda, W\}$
$\{f/V\}$	$\{-, W\}$	$\{f/\lambda, T\}$
$\{f/W\}$	$\{-, V\}$	$\{f/\lambda, U\}$
$\{f/T\}$	$\{-, U\}$	$\{f/\lambda, V\}$
$\{f/U\}$	$\{-, T\}$	$\{f/\lambda, W\}$

The associated PDA for the intersection follows.

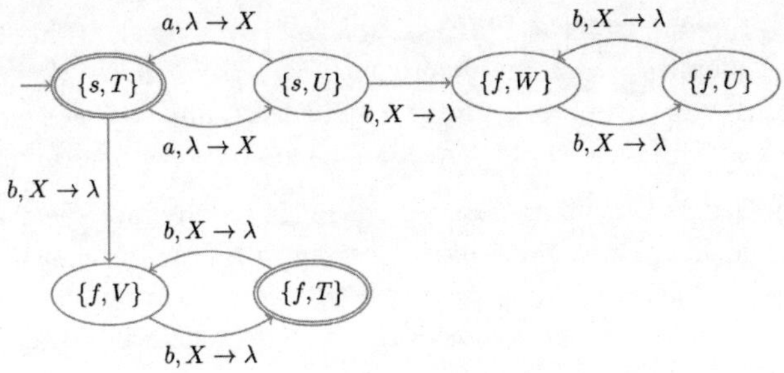

Figure 7–5: PDA for the regular intersection

Observe that states $\{f, W\}$ and $\{f, U\}$ constitute a composite jail, and can be removed:

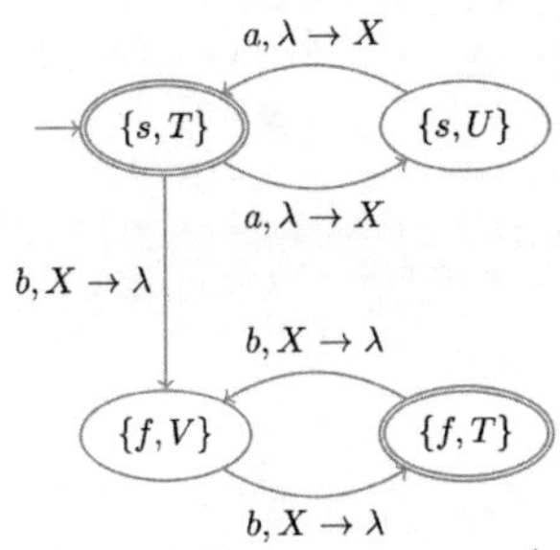

Figure 7–6: Final version of the PDA

CFLs are *not* closed under *complement.* This follows from the fact that they are closed under union but not under intersection. Using elementary set theory, we can express the intersection of sets as the complement of the union of the complements:

$$A \cap B \equiv \overline{(\overline{A} \cup \overline{B})}$$

If CFLs were closed under complement, then we could construct the intersection of any two CFLs, a *contradiction* to the fact that CFLs are not closed under intersection.

We may further conclude that CFLs are *not* closed under *difference* by a similar set-theoretic argument:

$$A - B \equiv A \cap \overline{B}$$

If the left-hand side above were context-free, the right-hand side would be also, but we know that, in general, it is not.

Closure Properties of DCFLs

As we consider the closure properties of deterministic context-free languages, it is important to keep in mind that we have made no connection between context-free grammars and determinism;

determinism for CFLs is defined in terms of PDAs. In fact, there is no algorithm to determine whether the language a CFG generates is deterministic context-free[1]. For a DCFL to be closed under an operation, the result must be a DCFL—i.e., it must *remain deterministic.* Some operations under which DFCLs are not closed may still yield context-free languages—just not deterministic ones.

In Chapter 5 we argued that DCFLs are not closed under union since there must be an arbitrary choice of which PDA to choose when accepting a string in the union of two languages. This does not necessarily mean that whenever we must choose between machines that the result will never be deterministic. It just means that not all unions of DCFLs will be deterministic.

The union of a DCFL and a *regular language* is deterministic, however. This follows from the fact that we can construct a product table for the union of a CFL with any regular language (which always has an associated deterministic automaton), and that product tables introduce no non-determinism. We can find a DPDA for the union of $L_1 = a^*$ and $L_2 = \{a^n b^n \mid n > 0\}$ without using a product table, however. The following DPDA recognizes $L_1 \cup L_2$. (We are using an end-of-string delimiter here.)

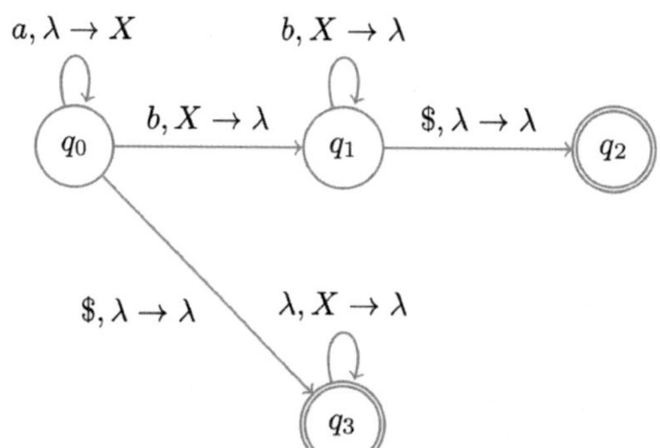

Figure 7–7: DPDA recognizing $a^* \cup a^n b^n$

This DPDA was easy to construct since L_1 is regular and therefore, does not use the stack. We must keep track of the a's in case there are b's that follow. If there aren't any b's, then we just need to empty the stack to accept the run of a's.

We also mentioned in Chapter 5 that DCFLs are closed under complement. By the set-theoretic identity for intersection in terms of union and complement (see two equations back), we may conclude that DCFLs are *not* closed under intersection, because we can't construct the union of two arbitrary DCFLs. They are closed under *regular intersection*, however, because, as with regular union mentioned above, intersecting a DPDA with a DFA by a product table does not introduce any non-determinism.

That leaves concatenation, Kleene star, and reversal.

DCFLs are not closed under concatenation in general because the suffix of the prefix (the part in the first language) of a concatenated string could match the prefix portion of the second language, and it would require non-determinism to make the switch and empty the stack before doing so. (The proof is complex and beyond the scope of this book.)

[1]There happens to be an *unambiguous CFG* for every DCFL, but that is a separate matter.

We may now conclude that DCFLs are not closed under Kleene star because that operation concatenates a language with itself, and DCFLs are not closed under concatenation.

We offer a counterexample to show that DCFLs are not closed under reversal. Recall the DCFL $\{wcw^R \mid w \in (a+b)^*\}$ from Figure 5–3. This is the language of odd length palindromes of a's and b's with a c in the middle. The following DPDA accepts the language $wcw^R(a+b)^*$.

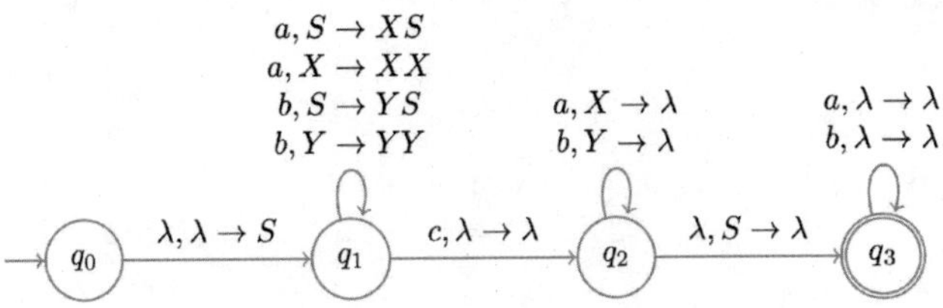

Figure 7–8: DPDA for $wcw^R(a+b)^*$

Now consider how to form a DPDA to accept the reverse language, $(a+b)^*w^Rcw$. For the $(a+b)^*$ prefix, we need to accept an arbitrary number of a's and b's *without using the stack*. The problem is, we don't know when the first a or b of w^R appears so that we can start using the stack—that must be done non-deterministically. But $wcw^R(a+b)^*$ is the concatenation of a DCFL with a regular language *on the right.* DCFLs are closed under right-regular concatenation, since the stack will be emptied before the regular suffix is processed, and regular languages do not use the stack.

DCFLs are *not* closed under union, intersection, concatenation, Kleene star, or reversal. They are *only* closed under **complement**, and under union, intersection, and right concatenation with *regular languages.*

Key Terms

regular intersection • regular union • right-regular concatenation

Exercises

1. Refer to the grammars in Examples 6–5 and 6–6 to find CFGs for the following:
   a. The union of the languages in Examples 6–5 and 6–6
   b. The concatenation of the languages in Example 6–5 and 6–6, in that order
   c. The Kleene star of the language in Example 6–5.

2. Find a PDA for the intersection of the PDA in Figure 5–1 (repeated below) and the language of strings of a's and b's where the numbers of a's is a multiple of 3 (you can figure out that DFA).

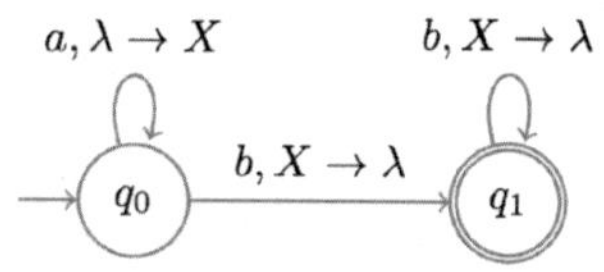

Copy of Figure 5–1

3. Are deterministic context-free languages closed under regular difference? Why or why not? (Note: The *regular difference* with a Context free language, C, and a Regular language, R, is C–R or R–C.)

7.3 Decision Algorithms

The most fundamental question to ask about a context-free language is whether a given string is in the language (this is the *parsing problem*). If a DPDA is available, then the answer is as simple as running a string through it, but many context-free languages are not deterministic. There happens to be an algorithm, however, that works for all context free languages. This is where Chomsky Normal Form for context-free grammars is useful.

Recall that rules in a CNF grammar come in only two forms: where the righthand side of a rule is either a single terminal, or exactly two non-terminals. The algorithm we have in mind begins by matching symbols of the string to the first type of rule, inferring the variables that produce each symbol, and then looks for rules where those variables appear as right-hand sides, so we can trace back to where those rules in turn originated. Continuing in this manner, if we eventually reach the start variable, then we know the string is in the language.

This algorithm, sometimes referred to as the "CYK Algorithm"[2], uses a problem-solving technique called *dynamic programming* to obtain the result. Dynamic programming solves problems in *stages*, and uses results from previous stages to solve later ones. We illustrate the algorithm by testing the string $baaa$ with the following CFG in Chomsky Normal Form:

$$
\begin{aligned}
S &\to XY \\
X &\to XA \mid a \mid b \\
Y &\to AY \mid a \\
A &\to a
\end{aligned}
$$

[2] In honor of the authors, Cocke, Younger, and Kasami, who discovered the algorithm independently.

We can easily see that this grammar does generate the string $baaa$. In fact, the grammar is ambiguous and there are multiple, distinct parse trees for this string. But let's get to the algorithm.

Stage 1

We begin by finding all the variables that can produce substrings of length 1, that is, that directly produce the terminal symbols a or b. For the grammar in question, we observe that b comes from X and a is generated by the variables X, Y, and A, which we denote as follows:

$$\begin{aligned} b &\leftarrow X \\ a &\leftarrow X, Y, A \end{aligned}$$

Stage 2

We now examine how all substrings of length 2 can be generated. There are only two distinct substrings of length 2 in the test string, namely ba and aa. We now use the information in Stage 1 to see where each component of each substring of length 2 comes from:

$$\begin{aligned} ba &\leftarrow X(X,Y,A) = XX, \underline{XY}, \underline{XA} \\ aa &\leftarrow (X,Y,A)(X,Y,A) = XX, \underline{XY}, \underline{XA}, YX, YY, YA, AX, \underline{AY}, AA \end{aligned}$$

The b in ba comes from the variable X, and the a comes from X, Y, or A. Therefore, the substring ba, if it can be generated by the grammar, must come from at least one of the concatenations of X with X, Y, or A, that is, either XX, XY, or XA. We now look at the grammar to see which of these pairs are valid, and in turn, which variables generates those pairs. Some variable pairings are not possible for this grammar (the valid pairings are underlined). We complete Stage 2 below, this time leaving out invalid variable concatenations.

$$\begin{aligned} ba &\leftarrow XY, XA \leftarrow \mathbf{S}, \mathbf{X} \\ aa &\leftarrow XY, XA, AY \leftarrow \mathbf{S}, \mathbf{X}, \mathbf{Y} \end{aligned}$$

This tells us that S and X can generate ba, and that S, X, and Y can generate aa. This information applies in the next stage.

Stage 3

The substrings of length 3 in $baaa$ are baa and aaa. Since we are using CNF, we know that variables only appear in pairs, so we look at all ways of partitioning these substrings into *two* components:

$$
\begin{aligned}
&baa:\\
&\qquad b\,aa \leftarrow X(S,X,Y) = XS, XX, \underline{XY} \leftarrow \mathbf{S}\\
&\qquad ba\,a \leftarrow (S,X)(X,Y,A) = SX, SY, SA, XX, \underline{XY}, \underline{XA} \leftarrow \mathbf{S,X}\\
&aaa:\\
&\qquad a\,aa \leftarrow (X,Y,A)(S,X,Y) = XS, XX, \underline{XY}, Y, YX, YY, AS, AX, \underline{AY} \leftarrow \mathbf{S,Y}\\
&\qquad aa\,a \leftarrow (S,X,Y)(X,Y,A) = SX, SY, SA, XX, \underline{XY}, \underline{XA}, YX, YY, YA \leftarrow \mathbf{S,X}
\end{aligned}
$$

Observe the use of previous stages. For the concatenation of b with aa, we use $b \leftarrow X$ from Stage 1 and $aa \leftarrow S, X, Y$ from Stage 2. We see that S and X can generate baa and S, X, and Y can generate aaa.

Stage 4

This final stage partitions the entire string into two concatenated substrings in three ways (we omit invalid variable pairings altogether this time for clarity):

$$
\begin{aligned}
&baaa:\\
&\qquad baa\,a \leftarrow (S,X)(X,Y,A) = XY, XA \leftarrow \mathbf{S,X}\\
&\qquad ba\,aa \leftarrow (S,X)(S,X,Y) = XY \leftarrow \mathbf{S}\\
&\qquad b\,aaa \leftarrow X(S,X,Y) = XY \leftarrow \mathbf{S}
\end{aligned}
$$

The goal in the last stage is to see if the start variable, S, appears. We could have stopped the procedure with the partition $baa\,a$ on the first line, answering "yes" to the decision question. If S is not obtained in the final stage, the grammar does not generate the string in question.

If you find this process cumbersome, there is a compact, tabular method that makes things a little easier to track. We first draw a stair-step diagram, positioning each symbol of the string in order by each corner "up the stairs". See Figure 7–9.

Diagonal (Stage) 1

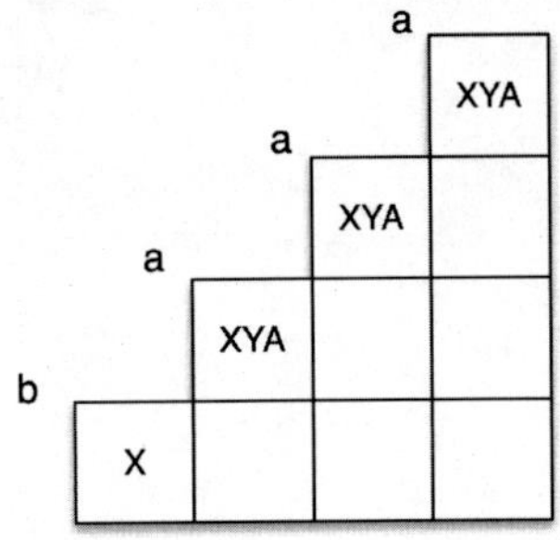

Figure 7–9: Recording Stage 1 of the CYK Algorithm

Each diagonal in the table represents a stage in the algorithm. Each box on the first diagonal holds the variables that generate the corresponding symbol, as we saw in Stage 1 earlier.

Diagonal (Stage) 2

For Stage 2, we first fill in the lower-left box in the second diagonal (the cell to the right of X and below XYA), then work our way up the diagonal. If you extend the southern line of that box all the way to the left, and the eastern line upward, they trace out the substring ba, which is how we began Stage 2 above. We then concatenate rules in the box for b with those in the first box for a, to obtain the concatenations $X(X, Y, A) = XX, XY, XA$. Continuing in this fashion, we fill the second diagonal to complete Stage 2 in the table:

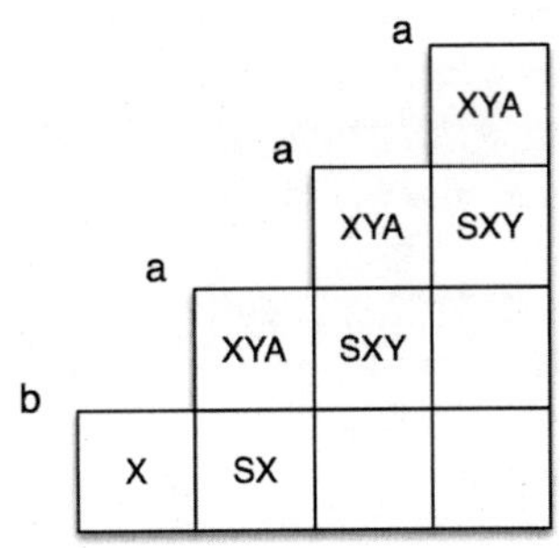

Figure 7–10: Recording Stage 2

Verify that you see how the table matches stages 1 and 2 previously.

To better visualize what is happening, we label each cell in the table according to its diagonal position, as follows.

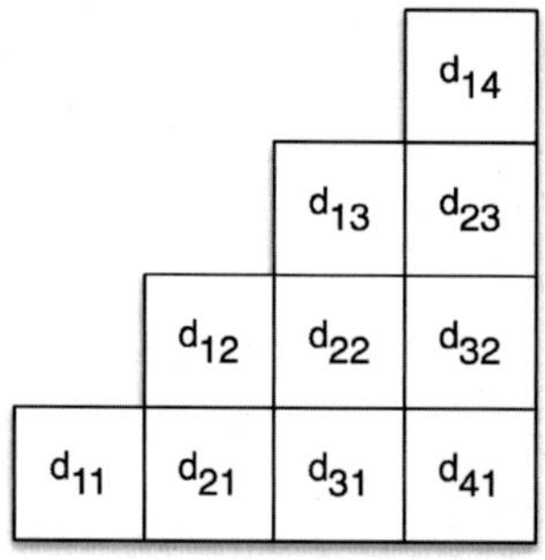

Figure 7–11: Labeling each cell diagonally

We can now represent the computation of Stage 2 as the following concatenations:

$$d_{21} = d_{11}d_{12}$$
$$d_{22} = d_{12}d_{13}$$
$$d_{23} = d_{13}d_{42}$$

We populated the second diagonal by concatenating the variables in the cell on the left of the target cell we are working on with those in the cell above and then finding the source of the resulting, valid variable pairs.

Diagonal 3

For Stage 3, we start with the substring baa, and therefore consider the concatenations "b aa" and "ba a". The corresponding variables in the table are the X under the b concatenated with the SXY in the second diagonal that traces the first aa, combined with concatenating the SX that generates ba with the XYA that generates an a. The key thing to notice about this pattern is that we concatenate variables from the cells on the *left* of the target diagonal cell with those *above* the target diagonal cell. The difference from Stage 2 is that there are *multiple pairs* of applicable boxes to concatenate. In general, to populate each cell on diagonal n, there are n–1 pairs of cells to concatenate and combine to fill in the new diagonal cell. The following diagram depicts which cells we need to concatenate to fill in d_{31} by giving them the same shading.

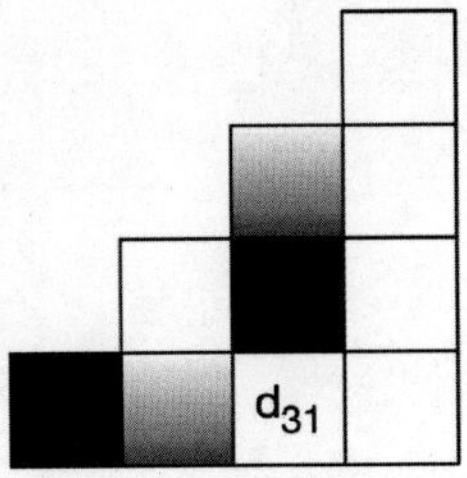

Figure 7–12: Filling in d_{31}

We populate the target cell d_{31} with the union of the variables that generate the pairs in our concatenations. The following diagram shows which cells combine for the other entry on the third diagonal (d_{32}, which corresponds to the substring aaa).

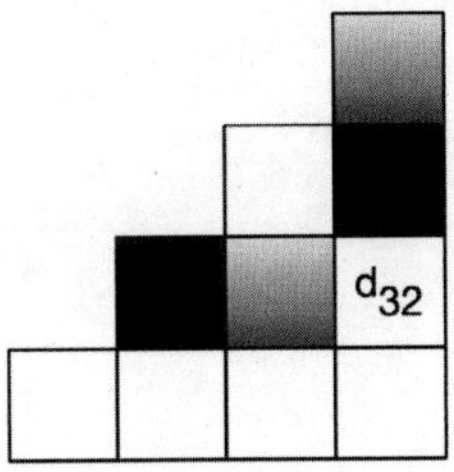

Figure 7–13: Filling in d_{32}

To confirm what we just did, the formulas for populating the diagonal for Stage 3 are as follows.

$$d_{31} = d_{11}d_{22} \cup d_{21}d_{13}$$
$$d_{32} = d_{12}d_{23} \cup d_{22}d_{14}$$

We now update our working CYK table:

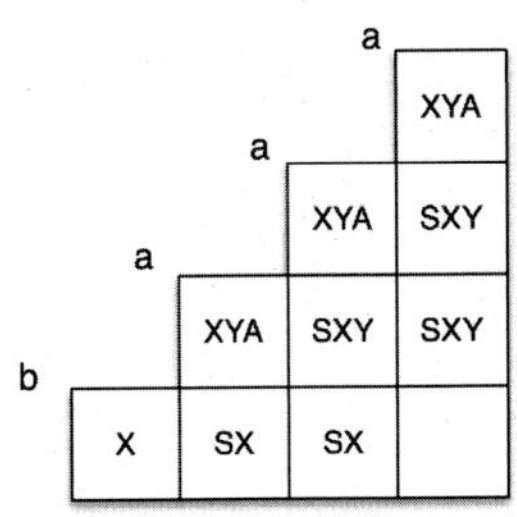

Figure 7–14: Completing stage 3

Diagonal 4

We formulate the final answer for Stage 4 according to the following pattern and formula.

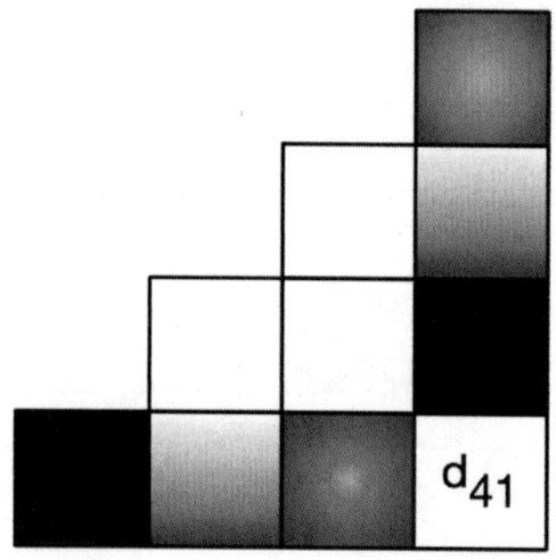

Figure 7–15: Filling in d_{41}

$$d_{41} = d_{11}d_{32} \cup d_{21}d_{23} \cup d_{31}d_{14}$$

Following this pattern, we obtain the same result we did manually at the beginning of this section. See Figure 7-16 below.

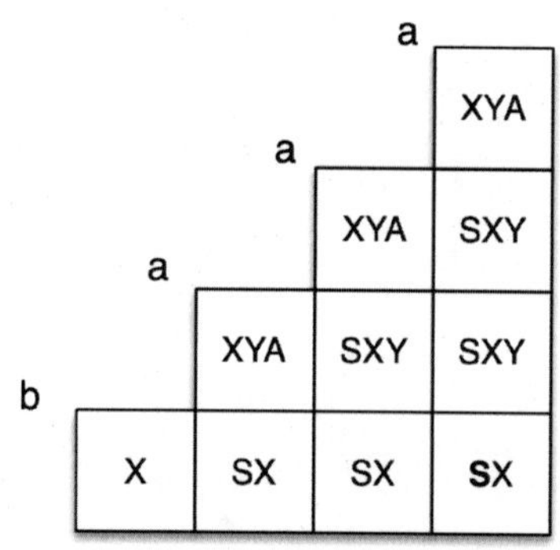

Figure 7–16: The completed CYK table

Since there is an S in the lower right cell, we know that the string *baaa* is generated by the grammar. Otherwise, we would conclude that it is not in the grammar's language.

Example 7–5

We now determine whether the following grammar generates the string *bbaa*.

$$
\begin{aligned}
&S \to AB \mid CD \mid a \mid b \\
&A \to a \\
&B \to SA \\
&C \to DS \\
&D \to a \mid b
\end{aligned}
$$

Since a comes from variables $[S, A, D]$, and b from $[S, D]$, the first diagonal representing the sources of the string $bbaa$ is $[SD, SD, SAD, SAD]$. See if you can carry out the steps that lead to the completed table below.

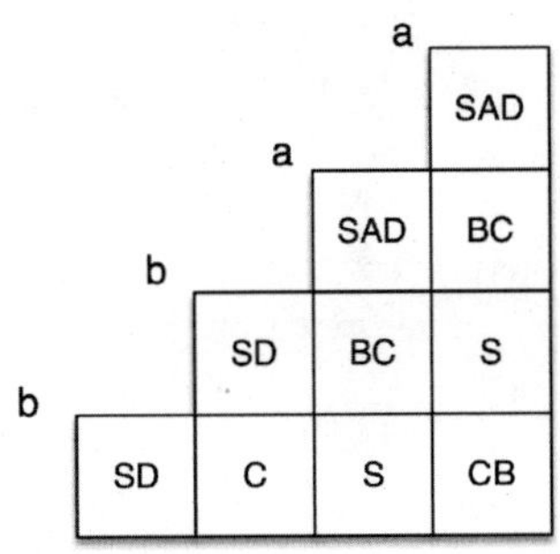

Figure 7–17: Completed CYK table for $bbaa$

Since the start variable S does not appear in the singleton cell in the last diagonal, this grammar does *not* generate the string $bbaa$.

The CYK algorithm is a "bottom-up" parsing algorithm because it begins with a string and backtracks using grammar rules to see if the string comes from the grammar's start variable. A "top-down" procedure (which we do not discuss in this book) begins with the start variable and tries to match rules with symbols of the string in a left-to-right traversal.

Is a CFL Empty or Infinite?

It is often easy to see if a CFG generates any strings at all by inspection, but, as always, we prefer an algorithm that will decide the question. The insight leading to such an algorithm is that all derivations end with "terminal rules"—rules that have only terminal symbols in them. A "bottom-up" approach suggests starting with a terminal rule and attempting to work our way back to the start variable. We will use the grammar below to illustrate the process.

$$S \to aA \mid bB \mid \lambda$$
$$A \to a \mid bBa \mid aAa$$
$$B \to b \mid aAb \mid bBb$$

It is obvious that this language contains the empty string, but let's look for a non-empty one. Using a copy of the grammar as a temporary workspace, we arbitrarily choose another terminal rule, $A \to a$, say, and substitute the terminal a for the variable A in every right-hand side rule:

$$S \to a\underline{a} \mid bB \mid \lambda$$
$$A \to a \mid bBa \mid a\underline{a}a$$
$$B \to b \mid a\underline{a}b \mid bBb$$

The goal is to see a terminal string on the start variable's right-hand side, so having obtained the expression $S \to aa$, we can stop the process and conclude that the string aa is in the grammar's language, therefore, the language has a non-empty string. Now that we know that this string is valid, we can derive it by reversing the process:

$$S \to aA \to aa$$

If a grammar generated no string, we would get to a point where we could do no further substitution and would be left with no terminal expressions on S's right-hand side.

Example 7–6

Does the following CFG generate any strings?

$$S \to XS$$
$$X \to YX \mid a$$
$$Y \to XX \mid YY$$

There is only one terminal rule, so we back-substitute throughout:

$$S \to \underline{a}S$$
$$X \to Y\underline{a} \mid a$$
$$Y \to \underline{aa} \mid YY$$

Another terminal rule has emerged, $Y \to aa$, which we propagate further:

$$S \to \underline{a}S$$
$$X \to \underline{aa} \mid a$$
$$Y \to \underline{aa} \mid \underline{aaaa}$$

No other substitutions are possible at this point, so this grammar generates no strings. We would have reached the same conclusion had we noticed that S is a non-terminating variable, of course, but this example affirms the validity of the algorithm.

To test whether a grammar generates an infinite language, first remove all useless variables. Having done so, all remaining variables contribute to forming a terminal string, so we can check whether any variable is ever *repeated* during a derivation. If so, that variable is (directly or indirectly) recursive, and therefore the language is infinite.

We begin by choosing which variable to check for recursion and "marking" it in every rule's right-hand side. Let's use the grammar at the beginning of this subsection (right before Example 7–6 above) and check the variable A (which is obviously recursive, as is the variable B). We first mark all A's on the right-hand side:

$$S \to a\underline{\mathbf{A}} \mid bB \mid \lambda$$
$$A \to a \mid bBa \mid a\underline{\mathbf{A}}a$$
$$B \to b \mid a\underline{\mathbf{A}}b \mid bBb$$

The next step is to mark every variable on the left-hand side that has a mark on its right-hand side:

$$\underline{\mathbf{S}} \to a\underline{\mathbf{A}} \mid bB \mid \lambda$$
$$\underline{\mathbf{A}} \to a \mid bBa \mid a\underline{\mathbf{A}}a$$
$$\underline{\mathbf{B}} \to b \mid a\underline{\mathbf{A}}b \mid bBb$$

All variables were marked. In particular, we marked the variable A, so that means we worked our way from A back to A, revealing that it is indeed a recursive variable. We conclude that the language is infinite. If A hadn't been recursive, we would try another variable. If no variables are recursive (directly or indirectly), the language is finite.

Example 7–7

Consider the following CFG.

$$S \to aA \mid SB$$
$$A \to baB \mid \lambda$$
$$B \to bB \mid bA$$

All variables are useful, and we see immediately that B is recursive, so the language is infinite. Nonetheless, let us apply the algorithm to see if A is also recursive. We first mark A on the right:

$$S \to a\underline{\mathbf{A}} \mid SB$$
$$A \to baB \mid \lambda$$
$$B \to bB \mid b\underline{\mathbf{A}}$$

Next, we mark all variables on the left that have a marked right-hand side rule:

$$\underline{\mathbf{S}} \to a\underline{\mathbf{A}} \mid SB$$
$$A \to baB \mid \lambda$$
$$\underline{\mathbf{B}} \to bB \mid b\underline{\mathbf{A}}$$

The variable A is not yet marked, but we see that we have reached the other variables, so we continue by marking S and B on the right:

$$\underline{\mathbf{S}} \to a\underline{\mathbf{A}} \mid \underline{\mathbf{S}}B$$
$$A \to ba\underline{\mathbf{B}} \mid \lambda$$
$$\underline{\mathbf{B}} \to b\underline{\mathbf{B}} \mid b\underline{\mathbf{A}}$$

We now see that A's right-hand side contains a marked variable, so we can mark A, showing that it also is a repeated variable.

Key Terms

parsing • dynamic programming • CYK algorithm

Exercises

1. Use the CYK algorithm to determine whether the grammar in Example 7–3 generates the string *bbab*. If it does, give a derivation.
2. Use the CYK algorithm to determine whether the grammar in Example 7–4 generates the string *abbb*. If it does, give a derivation.
3. Does the following grammar generate any strings?

$$
\begin{aligned}
S &\to XY \\
X &\to aX \\
A &\to a \\
Y &\to BY \mid BB \\
B &\to b
\end{aligned}
$$

4. Is the language of the following CFG finite or infinite?

$$
\begin{aligned}
S &\to AB \mid a \\
A &\to B \mid BB \mid BBB \mid b \mid c \\
B &\to a \mid ab \mid abc
\end{aligned}
$$

5. Is the language of the following CFG finite or infinite?

$$
\begin{aligned}
S &\to aA \mid b \\
A &\to AAb
\end{aligned}
$$

6. Is the language of the following CFG finite or infinite?

$$
\begin{aligned}
S &\rightarrow ABa \mid bAZ \mid b \\
A &\rightarrow Xb \mid bZA \\
B &\rightarrow bAA \\
X &\rightarrow aZa \mid bA \mid aaa \\
Z &\rightarrow ZAbA
\end{aligned}
$$

7.4 Infinite CFLs and Another Pumping Theorem

Now that we know how to determine if a context-free grammar generates an infinite language, let's look a little closer at such languages. Suppose A is a recursive variable in some CFG. Beginning with the start variable, until the time A first appears in a working derivation, symbols will have accumulated around A. Look at the CFG below and the derivation following.

$$
\begin{aligned}
S &\rightarrow aaB \\
A &\rightarrow bBb \mid \lambda \\
B &\rightarrow Aa
\end{aligned}
$$

$$S \Rightarrow aaB \Rightarrow aaAa$$

We will denote the aa prefix by u and the last a by z so we have $A \Rightarrow^* uAz$ so far. Now we continue substituting until A is repeated:

$$S \Rightarrow aaB \Rightarrow aaAa \Rightarrow aabBba \Rightarrow aabAaba$$

We have picked up $v = b$ and $y = ab$ around A, giving

$$S \Rightarrow^* uAz \Rightarrow^* uvAyz$$

We can make the same choices we just made between the occurrences of A to arrive at A again, giving

$$S \Rightarrow^* uAz \Rightarrow^* uvAyz \Rightarrow^* uvvAyyz = uv^2Ay^2z$$

In general, with any recursive variable, we can derive

$$S \Rightarrow^* uv^n Ay^n z$$

Eventually, we must terminate. We denote those terminating choices by $A \Rightarrow^* x$, finally arriving at $S \Rightarrow^* uv^n xy^n z$. In this instance, we can use the rule $A \Rightarrow \lambda$ and $n = 2$ ending with the string $uv^2xy^2z = aabbababa$, where $x = \lambda$. We could also go right to the terminal rules, avoiding generating v and y in the first place, by letting $n = 0$, which is the derivation

$$S \Rightarrow aaB \Rightarrow aaAa \Rightarrow aaa = uxz$$

Since B is also a recursive variable, we could use it as the "pivot" in a derivation, instead of A:

$$S \Rightarrow aaB \Rightarrow aaAa \Rightarrow aabBba \Rightarrow aabAaba \Rightarrow aab^2B(ba)^2 \Rightarrow aab^2Aa(ba)^2 \Rightarrow aabbababa$$

In this case $u = aa$, $v = b$, $x = a$, $y = ba$, $z = \lambda$. Once again, we could ignore u and v altogether and only derive $S \Rightarrow aaB \Rightarrow aaAa \Rightarrow aaa = uxz$.

For every infinite, context-free language, there are strings of the form $s = uv^n xy^n z \mid n \geq 0$.

The takeaway here is that there are portions of the derivation, namely v and y, that *repeat in parallel*. We will use this property to develop the Pumping Theorem for Context-Free Languages.

First, we must determine how long a string must be to force a repeated variable in a derivation, like we did with revisiting states in cycles in DFAs in Section 4.3. Once again, Chomsky Normal Form lends us a hand. Because rules of a CNF grammar are restricted, so are their associated derivation trees. We will relate the length of strings generated by a CNF grammar to the depth of their derivation trees.

The following CNF grammar generates the string *ababb*. Examine the derivation and tree that follow.

$$S \rightarrow AB$$
$$A \rightarrow BB \mid a$$
$$B \rightarrow SA \mid b$$

$$S \Rightarrow AB \Rightarrow BBB \Rightarrow SABB \Rightarrow ABABB \Rightarrow^* ababb$$

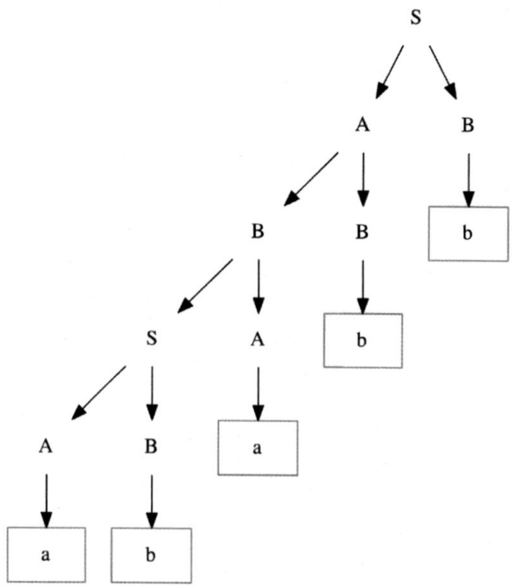

Figure 7–18: Derivation tree for ***ababb***

This tree has depth $d = 5$, but there are only three variables, so a variable must be repeated in some path. For example, both S and A are repeated in the leftmost path.

Since CNF rules only generate one terminal at a time, CNF derivation trees of depth $d = 1$ always have the following shape:

Figure 7–19: CNF derivation tree of depth 1

There is only one shape for CNF trees where $d = 2$ as well:

Figure 7–20: CNF derivation tree of depth 2

For $d = 3$, there are three possibilities:

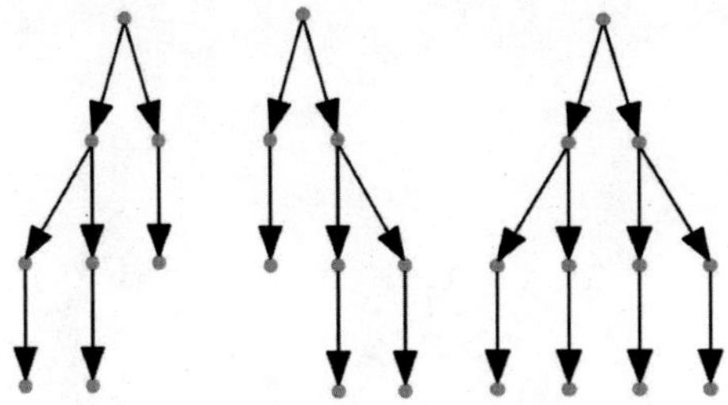

Figure 7–21: CNF derivation trees of depth 3

The second tree is the mirror image of the first. For $d = 4$, there are nine distinct trees, five of which appear below. The omitted trees are mirror images of the first four.

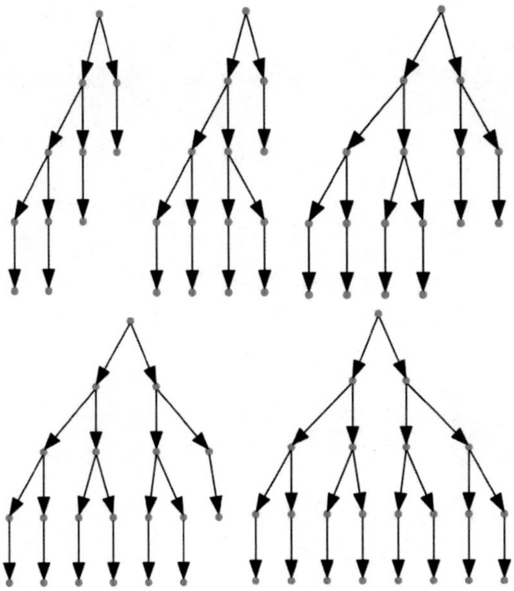

Figure 7–22: CNF derivation trees of depth 4

Now observe the following relationship between the depth of CNF derivation trees and the length of the strings they generate. See the table below.

Table 7–6: Relationship between tree depth and string length

Depth	Range of String Lengths
1	$[1, 1]$
2	$[2, 2]$
3	$[3, 4]$
4	$[4, 8]$
5	$[5, 16]$
d	$[d, 2^{d-1}]$

We now *invert* this relationship to express depth as a function of string length. From the last line in the table above we start from $l \leq 2^{d-1}$ and obtain the following:

$$l \leq 2^{d-1} \Rightarrow \log_2 l \leq d - 1 \Rightarrow d \geq \log_2 l + 1$$

So the depth is no less than one plus the logarithm of the string length.

If the depth of a tree is d, then it has a path with $d + 1$ nodes. Since the last element, the leaf, is a letter, there are d variables in that path. If we can force a repeated variable in that path, by choosing a string long enough, then we will be able to use the fact that the string will be of the form $s = uvxyz$, and then we can "pump" the derivation by repeating v and y together, obtaining strings of the form uv^nxy^nz. Let V be the set of variables in the CNF grammar. Then, to force a repeat (once again invoking the pigeonhole principle), we want at least $|V| + 1$ variables in the longest path, that is, we want[3] $d \geq |V| + 1$. Following this logic, and using the notation *lg* for the base-2 logarithm, we have

$$d \geq \lg l + 1 \geq |V| + 1 \Rightarrow \lg l \geq |V| \Rightarrow l \geq 2^{|V|}$$

So whenever a string is at least $2^{|V|}$ in length, where V is the set of variables in a CNF grammar, there will be a repeated variable in the longest path in the tree. Any such string can be partitioned as $s = uvxyz$, and can then be "pumped" so that $uv^nxy^nz \mid n \geq 0$ will also be in the language generated by the grammar. We are almost ready to state the Pumping Theorem for Context-Free Languages.

What's left to do is to find out something about the size of v and y, like we did with the condition $|xy| \leq p$ in the Pumping Theorem for regular languages. Let the string s be a string such that $|s| \geq 2^{|V|}$. By the discussion above we know that depth of the CNF parse tree, d, will satisfy $d \geq |V| + 1$. We also know that in this longest path of the tree there will be at least one repeated variable. We also know that a repeated variable allows us to partition the string as $s = uvxyz$.

We observe that in the longest path, going *up* from the terminal character there, there must be a repeated variable in an upward path of length no more than $|V| + 1$ (since there would be $|V| + 2$ nodes). The following diagram depicts such a variable, A.

[3]To have $|V| + 1$ variables, there will be $|V| + 2$ nodes, since the last is a leaf. The depth will be $|V| + 1$.

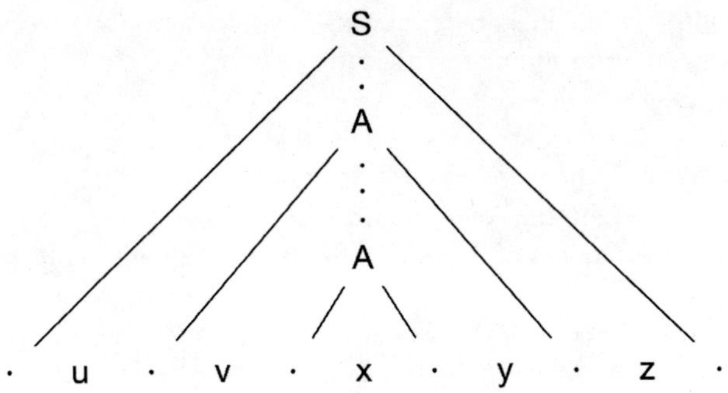

Figure 7–23: Tree schematic for $uvxyz$

The way we have chosen A guarantees that the subtree generated by the second occurrence of A (going up) is no greater than $|V|+1$ in depth. Per Table 7–6, a tree of depth $|V|+1$ can produce strings no longer than $2^{|V|}$. We now have our length condition for the pumping lemma: $|vxy| \leq 2^{|V|} \doteq p$.

Since there are no lambda or unit productions in a CNF grammar, then $|uv^iAy^iz| < |uv^{i+1}Ay^{i+1}z|$, where A is the repeated variable. In other words, at least one of v and y must be non-empty—that is, $|vy| > 0$.

We summarize the discussion above as a theorem.

The Pumping Theorem for Context Free Languages

For any infinite context-free language, L, there is a positive number, p, such that for every string, s, where $s \in L$ and $|s| \geq p$, s can be written as $s = uvxyz$ and the following hold:

1. $|vy| > 0$
2. $|vxy| \leq p$
3. $uv^ixy^iz \in L$, for $i \geq 0$

As with regular languages, we use this pumping theorem to show that a language is *not* context-free. The "rules of the game" are the same as for the Pumping Theorem for Regular Languages: we get to choose any qualifying string, but then must show that there is *no possibility* of satisfying this Pumping Theorem. As usual, if a language is "pumpable", we can draw no conclusions—the language may or may not be context-free.

Example 7–8

The language $a^nb^nc^n$ is not context-free. If it were, we could partition the string $a^pb^pc^p$ into five substrings, $uvxyz$, such that uv^ixv^iz also has runs of equal size of each letter. Since $|vxy| \leq p$, there are five cases to consider:

1. vxy appears wholly within the initial run of a's. In this case, as v and y are pumped, the number of a's increases while the number of the other letters doesn't change, failing the pumping theorem.
2. vxy contains both a's and b's. We may assume that $vxy = a^k b^m, m + k \leq p$. In this case, the number of a's and/or b's changes, but the c's do not change, again failing the pumping theorem.
3. vxy appears wholly within the run of b's. In this case, as v and y are pumped, the number of b's increases while the number of the other letters doesn't change, failing the pumping theorem, as in case 1 above.
4. vxy contains both b's and c's. As in case 2, the number of b's and/or c's changes, but the a's do not change, again failing the pumping theorem as in case 2 above.
5. vxy appears wholly within the run of c's. In this case, as in cases 1 and 3 above, as v and v are pumped, the number of c's increases while the number of the other letters doesn't change, failing the pumping theorem.

Example 7–9

The language of two equal halves, ww, is not context-free. Consider the string $a^p b^p a^p b^p$. Once again, since $vxy \leq p$, it is not possible for the various parts of the string to pump while keeping the halves of the string equal. There are seven cases:

1. vxy appears within the initial run of a's. In this case, as v and y are pumped, the number of initial a's increases while the number of a's in the third run doesn't change, failing the pumping theorem.
2. vxy contains both a's and b's from the first two runs. We may assume that $vxy = a^k b^m, m + k \leq p$. In this case, the number of a's and/or b's in the first half changes, but the two trailing runs of letters do not, again failing the pumping theorem.
3. vxy appears only within the first run of b's. In this case, the number of initial b's changes but the second run of b's does not, similar to case 1 above.
4. vxy contains b's followed by a's, failing in the same manner as case 2.
5. vxy is contained in the second run of a's, analogous to cases 1 and 3.
6. vxy contains a's and b's from the second half of the string, failing as in cases 2 and 4.
7. vxy is in the last run of b's, failing as cases 1, 3, and 5.

Example 7–10

The language $a^{n^2}, \Sigma = \{a\}$ is not context-free. Consider the string $s = a^{p^2} = uvxyz$. Since $|vxy| \leq p$, we know that $|vy| \leq p$. We will show that pumping up once to uv^2xy^2z does not reach the next square, $a^{(p+1)^2}$.

The difference between the two strings of square lengths is $(p+1)^2 - p^2 = 2p + 1$. However, the difference from pumping up once is

$$|uv^2xy^2z| - |uvxyz| = |vy| \leq p < 2p + 1$$

Pumping up once does not quite get to the next square length.

We could have pumped down once as well. The number of variables in a CNF grammar for an infinite language will be at least 2, and $p = 2^{|V|}$, so $p \geq 4$. Also, the distance down to the previous square is $p^2 - (p-1)^2 = 2p - 1$. We now have

$$|uvxyz| = |uxz| = vy \leq p < 2p - 1$$

Pumping down didn't make it far enough either.

In preparation for the next example, we need to show that if $n > 2$, then $n! - n > (n-1)!$. This can be verified graphically and is also illustrated by the Python program below.

Illustrating $n! - n > (n-1)! \mid n > 2$

```
import math

for i in range(3,10):
    print(math.factorial(i)-i, math.factorial(i-1))

''' Output:
3 2
20 6
115 24
714 120
5033 720
40312 5040
362871 40320
'''
```

This can also be proven by induction. For the base case we have $3! - 3 > 2! \equiv 3 > 2$. We now show that $n! - n > (n-1)! \rightarrow (n+1)! - (n+1) > n!$:

$$(n+1)! - (n+1) = (n+1)n! - n - 1 = n \cdot n! + n! - n - 1$$

We now apply the induction hypothesis to obtain

$$n \cdot n! + \mathbf{n! - n} - 1 > n \cdot n! + (\mathbf{n - 1})! - 1$$

The latter expression $n \cdot n! + (n-1)! - 1$ is clearly greater than $n!$.

Example 7–11

The language $a^{n!}, n > 2$ is not context-free. If it were, then for any $p > 2$ we could write $a^{p!} = uvxyz$, and then pump *down* once, subtracting v and y, and obtain a string of factorial length. Since, by the pumping theorem, $|vy| \leq p$, we have

$$|uvxyz| - |vy| = p! - |vy| \geq p! - p > (p-1)!$$

So, pumping down once doesn't make it all the way to $(p-1)!$, the next factorial down, failing the pumping theorem.

Example 7–12

The language $\{a^i b^j c^k \mid i < j \wedge i < k\}$ is not context free. The number of occurrences of a's in this language is strictly less than the number of the other letters. As we often do, we will pick a string which just barely qualifies: $a^p b^{p+1} c^{p+1}$.

Since $|vxy| \leq p$, the vxy part can contain:

1. only a's: pumping up increases the number of a's, violating the condition $i < j \wedge i < k$
2. a's followed by b's: pumping up breaks $i < k$
3. only b's: pumping down breaks $i < j$
4. b's followed by c's: pumping down breaks $i < j \wedge i < k$
5. only c's: pumping down breaks $i < k$

This language fails the Pumping Theorem for context-free languages

---

Key Terms

pumping theorem for context-free languages • pumping in parallel

Exercises

1. Show that the language $a^n b^m a^m, 0 \leq n \leq m$ is not context-free.
2. Show that the language a^n where n is prime is not context-free.

Chapter Summary

In this chapter, we showed how to convert any context-free grammar to Chomsky Normal Form, which in turn enabled us to illustrate the CYK algorithm for parsing strings, and to develop the Pumping Theorem for Context-free Languages. Remember that the Pumping Theorem can only be used to show that a language is *not* context-free, just like the Pumping Theorem for Regular Languages only shows that a language is not regular. Showing that a language can be pumped proves *nothing*.

The closure properties for deterministic vs. non-deterministic CFLs have little in common: only that *neither* is closed under intersection, and *both* are closed under regular intersection and regular union.

III Recursively Enumerable Languages

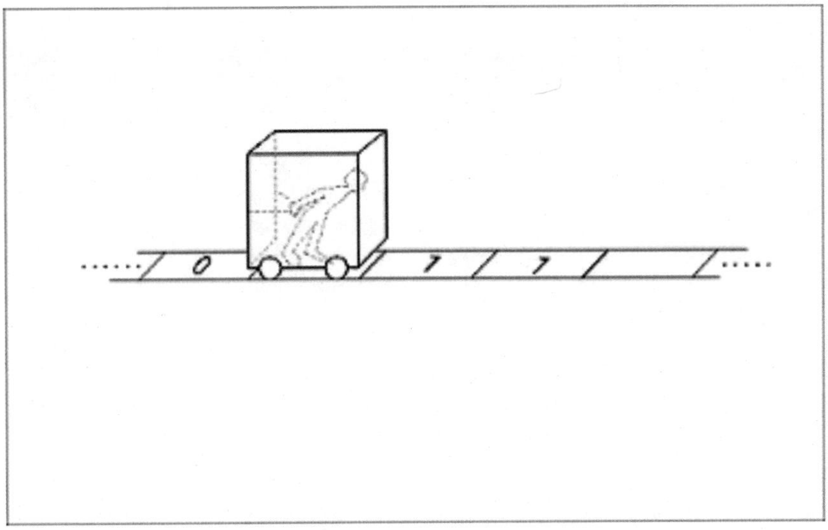

Diagram from Boolos, Burgess, and Jeffrey, Computability and Logic, Cambridge University Press, 2007, Figure 3-1, p. 25

8. Turing Machines

Unless in communicating with it one says exactly what one means, trouble is bound to result.

– Alan Turing

Where Are We?

Language Class	Recognized By	Generated By
Regular	Finite Automata	Regular Grammar Regular Expression
Context-Free	Pushdown Automata	Context-Free Grammar
Recursively Enumerable	**Turing Machine**	Unrestricted Grammar

Chapter Objectives

- *Become familiar with the operations and properties of Turing Machines*
- *Create Turing Machines for common computations and language-recognition problems*
- *Understand the Universal Turing Machine as a model for general-purpose computing*

The next, and final, type of machine we will study is the *Turing machine*, named after mathematician Alan Turing, who proposed the idea in 1936. At the time, he was trying to show that there could exist no algorithm to decide all logical mathematical questions, as the celebrated mathematician David Hilbert had proposed. It naturally followed that Turing needed to precisely define what an algorithm is. Others were also working on solving the same problem (notably Kurt Gödel, Alonzo Church, and

Stephen Kleene), but it was Turing's formulation of the nature of computations that proved to be the simplest and most effective, and also provided a sound theoretical foundation for the computer revolution.

His abstract "machines" prefigured what we now understand as a low-level, computer instruction set. More importantly, it was Turing that introduced the notion of a program reading other programs as input and simulating their behavior (i.e., "executing" them), which strongly influenced the development of the digital computer. But before we look at Turing machines, let's consider what we can do with a pushdown automaton with two stacks.

8.1 Prelude

We know that the language $a^nb^nc^n$ is not context-free, but a pushdown automaton with two stacks can recognize it. We use one stack to match the a's and b's, and the other to match the b's and c's. See the 2-stack PDA below.

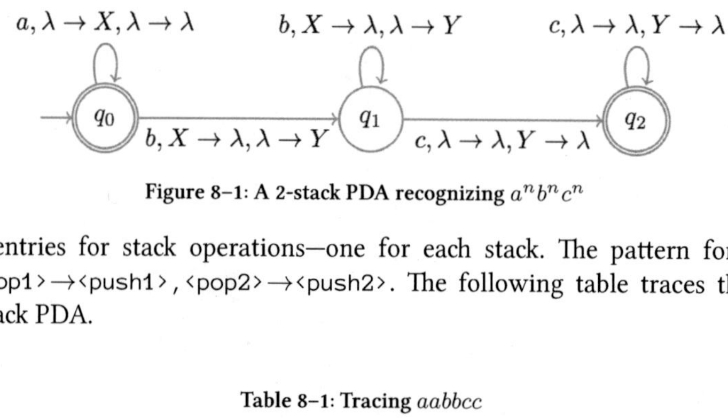

Figure 8–1: A 2-stack PDA recognizing $a^nb^nc^n$

There are two entries for stack operations—one for each stack. The pattern for each transition is <symbol>,<pop1>→<push1>,<pop2>→<push2>. The following table traces the string $aabbcc$ through the 2-stack PDA.

Table 8–1: Tracing $aabbcc$

State	Input	Stack 1	Stack 2
q_0	$aabbcc$	λ	λ
q_0	$abbcc$	X	λ
q_0	$bbcc$	XX	λ
q_1	bcc	X	X
q_1	cc	λ	XX
q_2	c	λ	X
q_2	λ	λ	λ

Now, what about the language $a^nb^nc^n$? Do we need another stack? No. We just move back to the first stack to match the c's and d's:

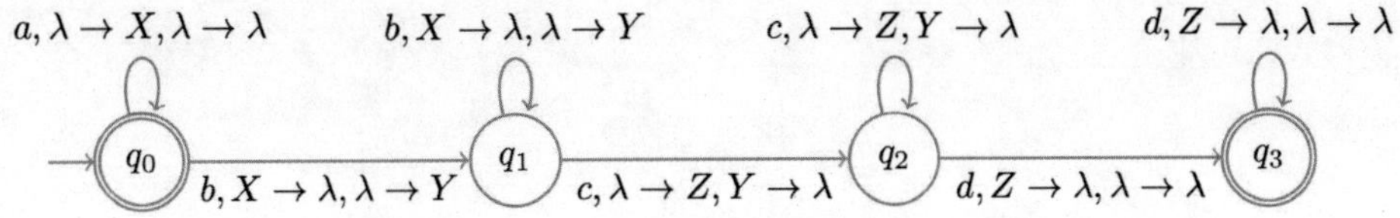

Figure 8–2: A 2-stack PDA recognizing $a^n b^n c^n d^n$

Here is a trace of *aabbccdd*:

Table 8–2: Tracing *aabbccdd*

State	Input	Stack 1	Stack 2
q_0	*aabbccdd*	λ	λ
q_0	*abbccdd*	X	λ
q_0	*bbccdd*	XX	λ
q_1	*bccdd*	X	X
q_1	*ccdd*	λ	XX
q_2	*cdd*	X	X
q_2	*dd*	XX	λ
q_3	*d*	X	λ
q_3	λ	λ	λ

A 2-stack PDA is equivalent to a Turing machine in computing power[1].

Post Machines

We obtain the same computational power from a Post[2] machine (aka "queue machine"), which is an automaton that uses a *queue* instead of two stacks. A single queue suffices to match a Turing machine's capability. By using an end-of-string marker, we can *cycle* through the input repeatedly by popping symbols from the front and pushing them onto the back of the queue. In addition, there is no separate "input channel", as we have been using—we assume that the queue is pre-populated with a \$-delimited input string before we begin processing. The following queue machine accepts the language $a^n b^n c^n$. As we often do, we have omitted handling invalid input (e.g., symbols out of order), in which case the machine crashes, implicitly rejecting the input string.

[1]A 2-stack PDA simulates a Turing machine by keeping everything to the left of the TM's read/write head on one stack, and everything else on the other stack. As the TM progresses, items are moved from one stack to the other, or removed, as needed.

[2]Named after logician Emil Post. His work preceded that of Turing and Church.

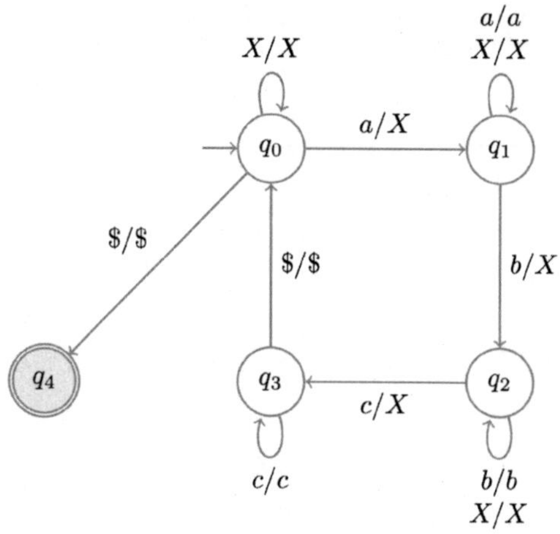

Figure 8–3: A Post (Queue) machine accepting $a^n b^n c^n$

The pattern for the edges is `<pop-front/push-back>`. We trace the action of the machine in the following table.

Table 8–3: Tracing $aabbcc$

State	Queue
q_0	aabbcc$
q_1	abbcc$a
q_1	bbcc$Xa
q_2	bcc$XaX
q_2	cc$XaXb
q_3	c$XaXbX
q_3	$XaXbXc
q_0	XaXbXc$
q_0	aXbXc$X
q_1	XbXc$XX
q_1	bXc$XXX
q_2	Xc$XXXX
q_2	c$XXXXX
q_3	$XXXXXX
q_0	XXXXXX$
q_0	XXXXX$X
q_0	XXXX$XX
q_0	XXX$XXX
q_0	XX$XXXX
q_0	X$XXXXX
q_0	$XXXXXX
q_4	XXXXXX$

The process replaces the first a it sees with an X, then looks for a b to replace, and then a c. When it reaches the end marker, it repeats the process. If it doesn't find an a, it checks that only X's remain. The machine crashes if that is not the case, rejecting the input string. The final state, q_4, is a *halt state* (there are no out edges). The entire input is on the queue, so processing terminates when a halt state is reached.

Key Terms

2-stack PDA • post/queue machine • halt state

Exercise

1. Draw a Post (queue) machine that accepts $n_a(w) = n_b(w) \mid w \in (a + b)^*$.

8.2 The Standard Turing Machine

Like the queue machine we just saw, a Turing Machine (TM) does not have a separate input channel besides its working storage. A Turing Machine is a state machine that has a *two-way infinite tape*, consisting of cells that hold a *single symbol.* We assume that the input is already written somewhere on the tape, and a read-write head is positioned at the first symbol of the input. The rest of the cells are "pre-formatted" with a special symbol called a "blank", which is not allowed to be part of the alphabet of any language.

The difference between a queue machine and a TM is how the data is accessed. Each step of a TM's computation reads the current cell, overwrites that cell (sometimes with the current contents, so there is no change in symbol), and then moves one cell either to the left or right. For this reason, we say that a TM has unrestricted memory: we can reach any cell we want by a series of moves. Since the tape is infinite in length, Turing machines also have unlimited memory. The following TM accepts the language $a^n b^n c^n \mid n > 0$. (*Note*: we use □ for blank. When drawing TMs by hand, it is convenient to use either B or an underscore for a blank, L for $\leftarrow$, and R for $\rightarrow$.)

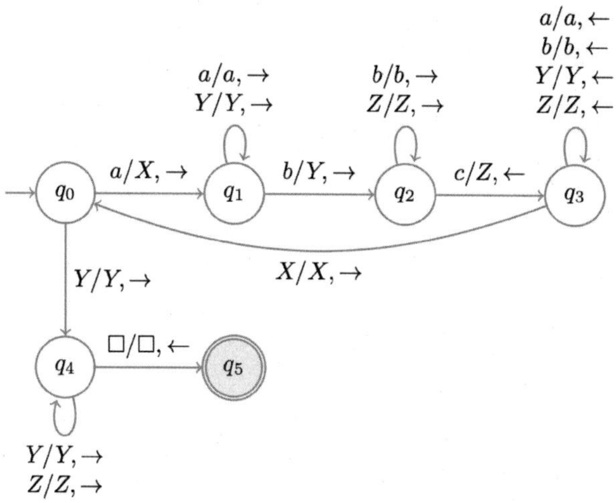

Figure 8–4: Turing machine accepting $a^n b^n c^n \mid n > 0$

The pattern for transitions in a Turing machine is `<read>/<write>,<direction>`. This machine marks each leading a, b, and c with X, Y, and Z, respectively, on each pass through the string. We don't need an end-of-string marker since blanks can serve that purpose. Also, to skip over a symbol, we merely write it over itself. Having no out-edges, q_5 is a halt state. Most of the time we don't include a non-accepting halt state because an invalid string will cause the machine to crash, and we consider a crash as a rejection. The trace of *aabbcc* appears below. (*Note*: the brackets are not actually present on the tape. They only serve to show under which symbol the read/write head is positioned.)

Table 8–4: TM trace of *aabbcc*

State	Tape Contents
q_0	`[a]abbcc`
q_1	`X[a]bbcc`
q_1	`Xa[b]bcc`
q_2	`XaY[b]cc`
q_2	`XaYb[c]c`
q_3	`XaY[b]Zc`
q_3	`Xa[Y]bZc`
q_3	`X[a]YbZc`
q_3	`[X]aYbZc`
q_0	`X[a]YbZc`
q_1	`XX[Y]bZc`
q_1	`XXY[b]Zc`
q_2	`XXYY[Z]c`
q_2	`XXYYZ[c]`
q_3	`XXYY[Z]Z`

Table 8–4: TM trace of $aabbcc$

State	Tape Contents
q_3	XXY[Y]ZZ
q_3	XX[Y]YZZ
q_3	X[X]YYZZ
q_0	XX[Y]YZZ
q_4	XXY[Y]ZZ
q_4	XXYY[Z]Z
q_4	XXYYZ[Z]
q_4	XXYYZZ[]
q_5	XXYYZ[Z]

Notice how there is always a blank at either end of the tape data when you need one (indicated by empty brackets in the table above). The TM halted with the read/write head positioned under the last Z.

An alternative notation for tracing a TM's execution that does not require a table places the current state in parentheses before the symbol referenced by the read/write head. Compare traversing the columns below to the trace in the previous table.

(q_0)aabbcc	(q_3)XxYbZc	XX(q_3)YYZZ
X(q_1)abbcc	X(q_0)aYbZc	X(q_3)XYYZZ
Xa(q_1)bbcc	XX(q_1)YbZc	XX(q_0)YYZZ
XaY(q_2)bcc	XXY(q_1)bZc	XXY(q_4)YZZ
XaYb(q_2)cc	XXYY(q_2)Zc	XXYY(q_4)ZZ
XaY(q_3)bZc	XXYYZ(q_2)c	XXYYZ(q_4)Z
Xa(q_3)YbZc	XXYY(q_3)ZZ	XXYYZZ(q_4)[]
X(q_3)aYbZc	XXY(q_3)YZZ	XXYYZ(q_5)Z

Here the state acts as a *cursor*, positioned immediately to the left of the symbol currently over the read/write head. Each of the strings above is called an *instantaneous description* (ID) of the machine, so a trace of a machine's execution is a sequence of ID strings.

Since TMs are more flexible than the other machines we have studied, you can think of them as a "low-level" notation for computation (think of TMs as the most fundamental machine "instruction set" possible). For this reason, TMs are generally more complicated than simpler automata, just like using assembly language is more tedious than using a high-level programming language. The 12-state TM in Figure 8–5 below accepts the language of an equal number of a's, b's, and c's.

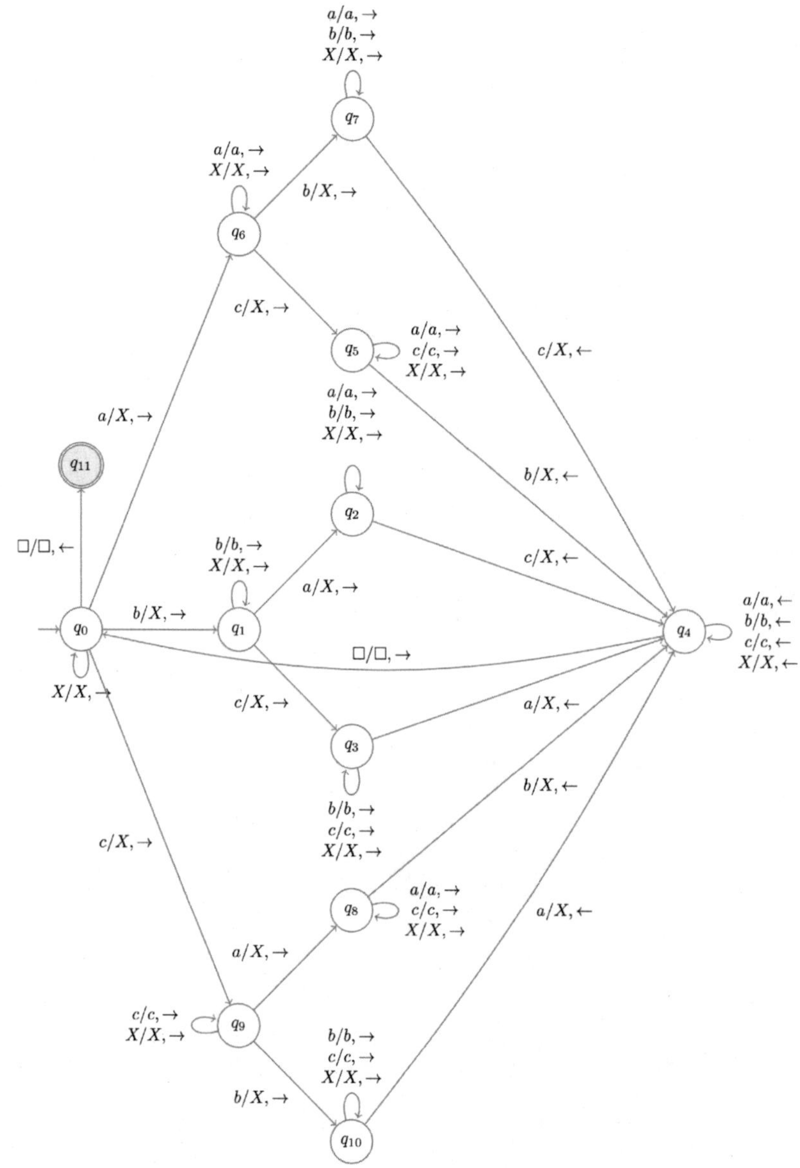

Figure 8–5: TM accepting $n_a(w) = n_b(w) = n_c(w) \mid w \in (a+b+c)^*$

This TM must account for the fact that the symbols can occur in any order. It looks a little intimidating, but it is just a three-way branch on the first of the three distinct symbols encountered. If an a occurs first, then we take the top branch, looking for a b followed by a c, or a c followed by a b. The other branches work analogously. We mark tracked symbols with an X. The first half of the trace of $baabcc$ follows. To save space we use the turnstile symbol to separate the IDs.

$$
\begin{aligned}
(q_0)baabcc &\vdash X(q_1)aabcc \vdash XX(q_2)abcc \vdash XXa(q_2)bcc \vdash XXab(q_2)cc \\
&\vdash XXabX(q_4)c \vdash XXab(q_4)Xc \vdash XXa(q_4)bXc \vdash XX(q_4)abXc \\
&\vdash X(q_4)XabXc \vdash (q_4)XXabXc \vdash (q_4)\Box XXabXc \vdash \Box(q_0)XXabXc \cdots
\end{aligned}
$$

This machine marks the first occurrence of each symbol, then rolls all the way back to the left to repeat the process. If the counts are unequal, the machine crashes (i.e., has no option to move), implicitly rejecting the string. The following pseudocode summarizes this TM's behavior.

Pseudocode for the TM in Figure 8–5

```
repeat
    skip all occurrences of X
    if a blank is encountered, ACCEPT
    if the current symbol is 'a':
        mark it with an X
        move right until, either a 'b' or 'c' is found
        if a 'b' was found:
            overwrite it with an X and move right until a 'c' is found
            overwrite the 'c' with an X
        else if a 'c' was found:
            overwrite it with an X and move right until a 'b' is found
            overwrite the 'b' with an X
    else if the current symbol is 'b':
        mark it with an X
        move right until, either a 'a' or 'c' is found
        if a 'a' was found:
            overwrite it with an X and move right until a 'c' is found
            overwrite the 'c' with an X
        else if a 'c' was found:
            overwrite it with an X and move right until a 'a' is found
            overwrite the 'a' with an X
    else if the current symbol is 'c':
        mark it with an X
        move right until, either a 'a' or 'b' is found
        if a 'a' was found:
            overwrite it with an X and move right until a 'b' is found
            overwrite the 'b' with an X
        else if a 'b' was found:
            overwrite it with an X and move right until a 'c' is found
            overwrite the 'c' with an X
    move left until the beginning of the input
```

As with all programming, it is helpful to design the logical steps of a solution before "coding".

Example 8–1

Turing machines can also act as *functions* by leaving output on the tape. The following Turing machine computes $n \bmod 2$. We encode numbers in *unary* notation, so $1 = 1, 2 = 11, 3 = 111$, etc. The idea is to use two states for evenness vs. oddness, and then leave either a 0 or 1 on the tape.

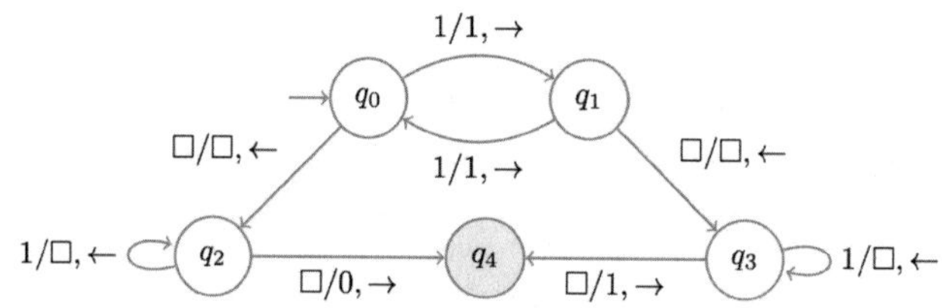

Figure 8–6: TM that computes n mod 2

States q_2 and q_3 change all 1's to blanks. q_4 is the halt state. When a number is even, a blank will eventually appear while in state q_0, so we erase all the 1's and leave a 0 on the tape. From q_1 we end up leaving a 1. There is no need for an accepting state since the machine produces output.

Example 8–2

It is easy to add numbers in unary notation. We separate the input numbers with a zero (or any symbol besides 1) and then just shuffle all the 1's left one position. For example, if we want to add $3 + 4$, then we begin with `11101111` on the tape and end up with `1111111`. The following TM computes $x + y$.

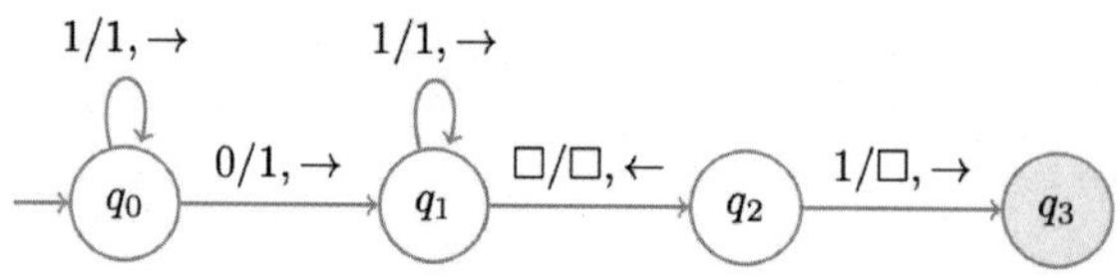

Figure 8–7: TM that computes $x + y$

The first state skips 1's until it finds a 0. The transition to q_1 overwrites the 0 with a 1. State q_1 skips the rest of the 1's. Finally, the last 1 is replaced by a blank, leaving $x + y$ ones on the tape.

Example 8–3

The following TM rounds its unary input up to the next multiple of 3 by appending one or two 1's as needed.

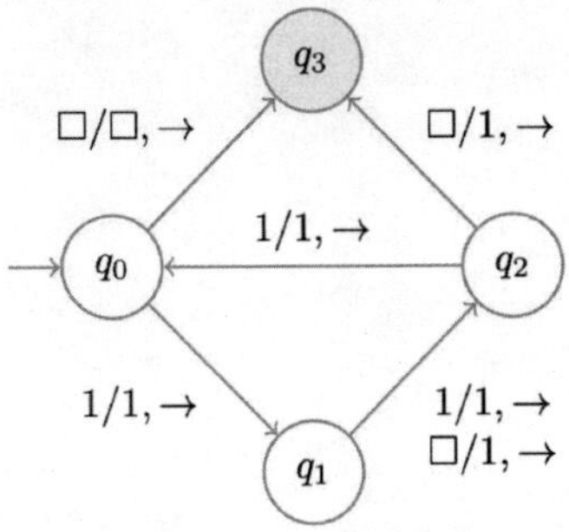

Figure 8–8: Rounds up to the next multiple of 3

This machine assumes that an empty tape represents the number zero (the symbol 0 is not used). Observe how state q_1 takes advantage of state q_2 when adding ones at the end.

Example 8–4

How would we compute n div 2 (aka $\lfloor n/2 \rfloor$)? We can look upon a unary number as *pairs* of 1's, possibly with one symbol left over. Dividing by 2 is keeping half of each pair. Suppose the number is $1111 (= 4)$. We can consider the first and last 1's a *pair*: we'll erase the first and mark the last, giving $11X$. Now we can look at the remaining pair of 1's and erase one and mark the other, leaving XX. When we run out of 1's, we can change the X's back to 1's, leaving us with $11 = 2 = 4$ div 2.

Suppose that the number is odd instead: 111, say. We begin the same way, erasing the first and marking the last: $1X$. Repeating the process as before, we erase the remaining 1, but there is no matching 1 for it. We then change all the X's to 1's, leaving 1. This scheme automatically rounds down for odd numbers. See Figure 8–9. The traces for 111 and 1111 follow.

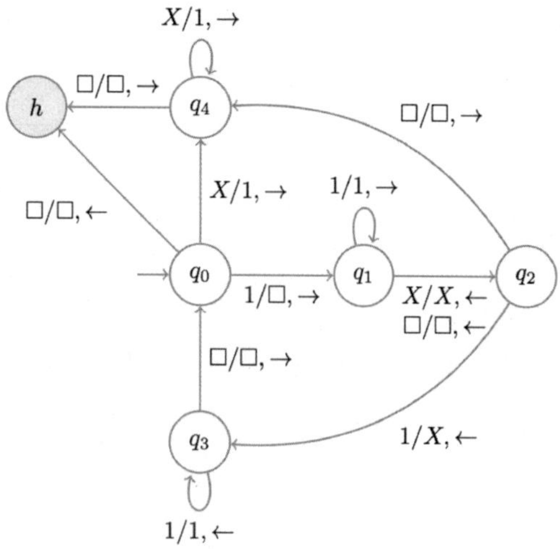

Figure 8–9: Computes n div 2

$$\begin{aligned}(q_0)111 &\vdash (q_1)11 \vdash 1(q_1)1 \vdash 11(q_1)\square \vdash 1(q_2)1 \vdash (q_1)1X \\ &\vdash (q_3)\square 1X \vdash (q_0)1X \vdash (q_1)X \vdash (q_2)\square X \\ &\vdash (q_4)X \vdash 1(q_4)\square \vdash (h)1\end{aligned}$$

$$\begin{aligned}(q_0)1111 &\vdash (q_1)111 \vdash 1(q_1)11 \vdash 11(q_1)1 \vdash 111(q_1)\square \vdash 11(q_2)1 \\ &\vdash 1(q_3)1X \vdash (q_3)11X \vdash (q_3)\square 11X \vdash (q_0)11X \vdash (q_1)1X \\ &\vdash 1(q_1)X \vdash (q_2)1X \vdash (q_3)\square XX \vdash (q_0)XX\square \vdash 1(q_4)X \\ &\vdash 11(q_4)\square \vdash 1(h)1\end{aligned}$$

If the input is 1 or blank, the tape is left blank to indicate a 0 result.

Perhaps you have noticed that in this chapter we have not used the term “character” or “letter”—we have spoken of “symbols” instead. This is intentional. The following example suggests why.

Example 8–5

We now design a Turing machine that adds binary numbers. For this example, we will assume that the numbers are given as pairs of bits in normal left-to-right order. Let’s calculate $4 + 13 = 17$ in binary:

0100

<u>1101</u>

10001

Adding 4 + 13 in binary

For simplicity, we "preprocess" the input by prepending a leading zero to the smaller number, so that the two numbers have the same number of bits. The sequence of bit-pairs to process is therefore (0,1),(1,1),(0,0), and (0,1). We will consider each of these pairs a "symbol". If you are uncomfortable with this, you can use variables to stand for each of these pairs (see section 8.3), but it is not necessary. The definition of a symbol is intentionally quite liberal for TMs. The first thing we will do is move to the right boundary and work our way backwards, just like we add numbers by hand. We will also use a separate state (q_2) to know when we have a carry bit. See the diagram below.

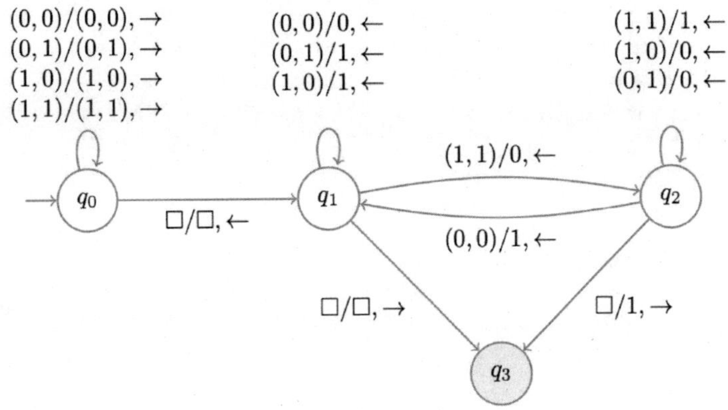

Figure 8–10: Binary adder

Here is a trace for adding $4 + 13$:

$$
\begin{aligned}
(\mathbf{q_0})(0,1)(1,1)(0,0)(0,1) &\vdash (0,1)(\mathbf{q_0})(1,1)(0,0)(0,1) \vdash (0,1)(1,1)(\mathbf{q_0})(0,0)(0,1) \\
&\vdash (0,1)(1,1)(0,0)(\mathbf{q_0})(0,1) \vdash (0,1)(1,1)(0,0)(0,1)(\mathbf{q_0})\square \\
&\vdash (0,1)(1,1)(0,0)(\mathbf{q_1})(0,1) \vdash (0,1)(1,1)(\mathbf{q_1})(0,0)1 \\
&\vdash (0,1)(\mathbf{q_1})(1,1)01 \vdash (\mathbf{q_2})(0,1)001 \\
&\vdash (\mathbf{q_2})\square 0001 \vdash 1(\mathbf{q_3})0001
\end{aligned}
$$

We now formally define the Turing Machine.

Definition 8.1

A standard **Turing Machine** is a *deterministic* automaton with a two-way, infinite tape along with the following:

- A set of states, Q, one of which is the initial state, and a non-empty subset of which, H, say, contains **halt states**
- An **input alphabet**, Σ, of language symbols
- A **tape alphabet**, Γ, which includes Σ, a special "blank symbol", $\Box$, and other arbitrary symbols as needed
- A **transition function**, $\delta : (Q - H) \times \Gamma \to Q \times \Gamma \times \{\leftarrow, \rightarrow\}$

Halt states have no out edges, hence the expression $(Q - H)$ as the domain for δ. Turing machines typically have only one or two halt states (a second halt state can be used to explicitly reject input). δ is sometimes called a **program**. The formal notation for the TM in Figure 8–8 above is as follows:

$$Q = \{q_0, q_1, q_2, q_3\}, \text{ with } q_3 \text{ as the halt state}$$
$$\Sigma = \{1\}$$
$$\Gamma = \{1, \Box\}$$
$$\delta(q_0, 1) = (q_1, 1, \rightarrow), \delta(q_0, \Box) = (q_3, \Box, \rightarrow), \delta(q_1, 1) = (q_2, 1, \rightarrow),$$
$$\delta(q_1, \Box) = (q_2, 1, \rightarrow), \delta(q_2, 1) = (q_0, 1, \rightarrow), \delta(q2, \Box) = (q_3, 1, \rightarrow)$$

Subroutines

We have included many practical examples of Turing machines to give the impression that using a TM is like using a programming language on a computer. Indeed, TMs are the conceptual basis for the design of computer algorithms as well as for computers themselves. We now show how to use TMs as *subroutines*. As with assembly language, it's mostly a matter of putting results in a format and location that supports use by other TMs. The call-and-return mechanism consists of joining components with transitions.

A useful subroutine copies tape contents from one place to another. We will design a TM that copies strings of a's and b's by appending its input to the right of the populated part of the tape. We will delimit the original input with a trailing # symbol. The plan is to mark each a with X and b with Y. We then traverse each marked symbol left-to-right, appending what it stands for to the right end of the used part of the tape, and "unmark" each X or Y by reverting to its respective original a or b. The following trace shows our plan with input *abbab#*:

$$abbab\# \vdash^* XYYXY\# \vdash^* aYYXY\#a \vdash^* abYXY\#ab$$
$$\vdash^* abbXY\#abb \vdash^* abbaY\#abba \vdash^* abbab\#abbab$$

Here is the Turing machine:

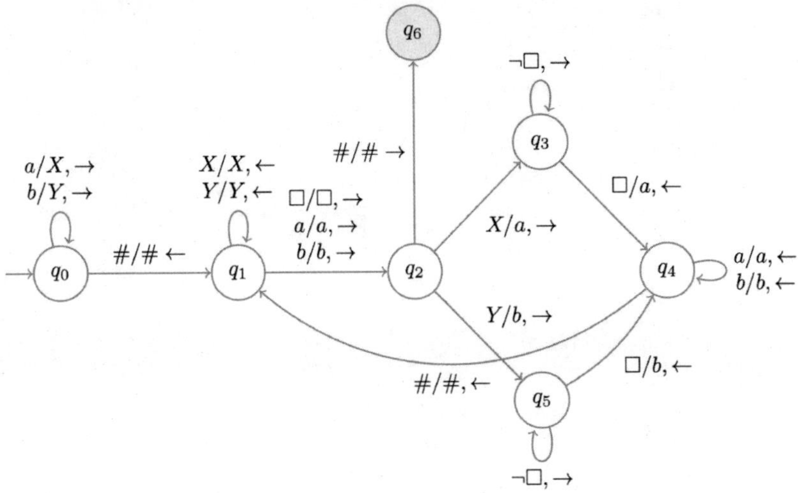

Figure 8–11: Copies strings of a's and b's

We introduced a shorthand on states q_3 and q_5. The expression "$\neg\Box, \rightarrow$" means "skip everything moving right until a blank is encountered". In this case, it is shorthand for

$$X/X, \rightarrow, Y/Y, \rightarrow, \#/\#, \rightarrow, a/a, \rightarrow, b/b, \rightarrow$$

Here is a trace for input $ab\#$:

$$(q_0)ab\# \vdash X(q_0)b\# \vdash XY(q_0)\# \vdash X(q_1)Y\# \vdash (q_1)XY\# \vdash (q_1)\Box XY\# \vdash (q_2)XY\#$$
$$\vdash a(q_3)Y\# \vdash aY(q_3)\# \vdash aY\#(q_3)\Box \vdash aY(q_4)\#a \vdash a(q_1)Y\#a \vdash (q_1)aY\#a$$
$$\vdash a(q_2)Y\#a \vdash ab(q_5)\#a \vdash ab\#(q_5)a \vdash ab\#a(q_5)\Box \vdash ab\#(q_4)ab \vdash ab(q_4)\#ab$$
$$\vdash a(q_1)b\#ab \vdash ab(q_2)\#ab \vdash ab\#(q_6)ab$$

If we were to run the machine again with configuration $(q_0)ab\#ab$, the output would be $ab\#(q_6)abab$, having appended a second copy of ab. Whenever we want to copy a string of a's and b's, we ensure that the string is #-delimited, position the read/write head under the first symbol of the string to be copied (which must be preceded by a blank) and have a transition connect from there to the start state of the copy subroutine.

Example 8–6

We now implement a TM that multiplies two unary numbers. The strategy is simple: for every 1 in the first number, we call an appropriate subroutine to append a copy of the second number to the end of the used portion of the tape. To compute 3×4, we begin with the following tape contents:

111□1111#

and end with twelve 1's appended to the original input:

111□1111#111111111111

For each 1 in the first number (x), we position the read/write head under the first 1 in the second number (Y) and "call" the following copy-TM (Figure 8–12 below) to append the 1's in y. It is like the machine in Figure 8–11 except we only have a single symbol to deal with (1, instead of a and b).

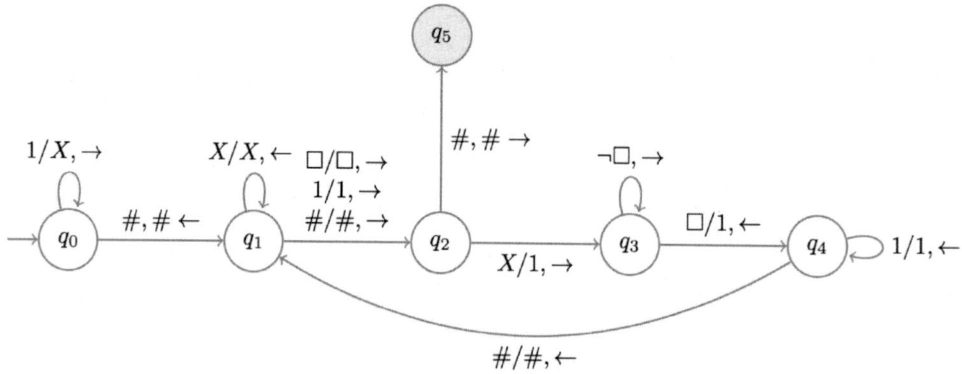

Figure 8–12: A COPY TM (Copies 1's)

This TM leaves the read/write head under the first symbol of the result string, which our "multiply" TM will account for when the COPY-TM halts. The plan is to mark the first 1 in x with an a, and then move to the beginning of y and call the COPY routine. We repeat this on each 1 in x. When we are finished, we restore the a's in x to 1's. See Figure 8–13 below.

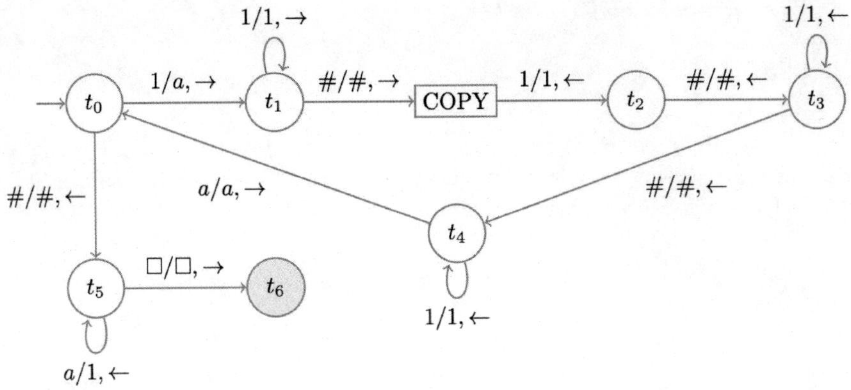

Figure 8–13: Multiples two unary numbers by copying 1's

To understand how this TM works, visualize that the transition exiting t_1 connects to state q_0 of Figure 8–12, and the edge into t_2 above comes from the COPY-TM's halt state, q_5. This is reminiscent of how we use the stack in assembly languages to facilitate communication between calling and called routines. An abbreviated trace appears below:

$$
\begin{aligned}
(t_0)111\square1111\# &\vdash^* a(t_0)11\square1111\#1111 \vdash^* aa(t_0)1\square1111\#11111111 \\
&\vdash^* aaa(t_0)\square1111\#111111111111 \vdash^* (t_6)111\square1111\#111111111111
\end{aligned}
$$

Halting

Whether a TM halts or not is a feature of the machine itself and its input. We emphasize the notion of halting because *some TMs do not halt on some inputs.* This is not a matter of "bad programming", but rather of the nature of certain computations. To illustrate, consider an arbitrary function $f : Z \times Z \to Z$, that takes two integer parameters and returns an integer. Suppose further that we are interested in the smallest non-negative integer, p, for which $f(x, p) = 1$ for some given x. Since f is arbitrary, we have no choice but to attempt to "solve" this problem by creating a brute-force function, like the function$g$ in the following Python code.

```
def g(f,x):
    p = 0
    while f(x,p) != 0:
        p += 1
    return p
```

Since f is arbitrary, it is possible that it never returns 1 for some choices of x. In such cases g will loop forever (or "hang"). This turns out to be related to a central issue in the theory of computation, which we will cover in some detail in the next two chapters.

Some Turing machines do not halt on some inputs

This is the price we pay for such a flexible, powerful model of computation as the Turing machine. A Turing machine that halts on every input constitutes an *algorithm*. A function for which a TM exists that always halts is called a **total function**, while a function that fails (i.e., does not halt) on some inputs (like $f(x) = \frac{1}{x}$, which fails for $x = 0$) is called a **partial function**. A language for which there is a TM that always halts, either rejecting or accepting its input, is a **recursive language** (aka a *Turing-decidable* language). A language that is *recognized* by a TM that *does not halt* on every input is called a **recursively enumerable language** (or a *Turing–recognizable language*). We will say more about the connection between languages and Turing machines in the next chapter.

At about the same time Alan Turing developed his machines, the mathematician Alonzo Church was also developing an alternative model of computation, the *lambda calculus*, which uses the mathematical notion of a function to perform any symbolic computation. The lambda calculus has the same computing power as Turing machines and forms the basis for functional programming languages such as Lisp, Scheme, ML, and Haskell.

In 1966, researchers Bohm and Jacopini showed that a programming language that supports the following three constructs is equivalent in computing ability to a Turing machine:

1. Stored computations performed in **sequence** (like statements in a programming language)
2. **Selecting** among two computations based on a Boolean expressions (like `if-then-else` statements)
3. **Repeating** a computation until a Boolean expression changes (loops)

These abilities define the discipline of structured programming, which has been a dominant programming style since the 1970s. This means that if we want to reason about computation in general, we can use simple, structured programs, like the Python examples in this book, instead of Turing machines. These results are the very foundation of computer science and are succinctly expressed in the Church-Turing Thesis:

Church-Turing Thesis

Any computation that can be carried out by mechanical means can be performed by some Turing machine.

This is a "thesis" because it can't be proven, but no one has yet found a solvable problem that can't be solved by a Turing machine[3].

[3]For a proof that TMs can simulate modern digital computers, see Rich, E., *Automata, Computability, and Complexity: Theory and Applications*, Prentice-Hall, 2008, pp. 393–397.

Key Terms

standard Turing machine • instantaneous description (ID) • subroutines • halting • structured programming • Church-Turing thesis

Exercises

1. Construct a Turing machine that adds two unary numbers, as in Example 8–2, except that it erases the first 1 and changes the separating zero to a 1.
2. Draw a Turing machine that takes a string representing two unary numbers, x and y, separated by a 0, and determines whether $x \geq y$. For example, the input for $x = 3, y = 4$ would be 11101111. Use two halt states: one for yes and one for no.
3. Give the trace of your machine in the previous problem processing the strings 11101111 and 11110111.
4. Draw a TM that computes $f(w) = w^R \mid w \in (a+b)^*$. For simplicity, you can use a delimiter and just copy the letters backwards at the right, so, for example, *abb#* becomes *abb#bba*.
5. Draw a TM that accepts unary numbers representing powers of 2. (Hint: Except for 1 and 2, dividing powers of 2 by 2 always gives an even number.)

8.3 Variations on Turing Machines

The machine that added binary numbers in Example 8–5 allowed pairs of bits as input symbols. Another way to look at that example is as a machine with a 2-track tape. See the tape in the diagram below.

...		0	1	0	0		...
...		1	1	0	1		...

Figure 8–14: A 2-track tape

We will design a TM that adds these numbers together, column-wise as before, leaving the tape with the following contents, the sum in the bottom track and the top track all blank.

…								…
…		1	0	0	0	1		…

Figure 8–15: The sum on a 2-track tape

As we did in Example 8–5, we will show that a standard, 1-track Turing machine can simulate a 2-track machine by introducing new symbols to stand for pairs of symbols, as follows.

$$
\begin{aligned}
X &= (0,0)\\
Y &= (0,1)\\
Z &= (1,0)\\
W &= (1,1)\\
\square &= (\square,\square)\\
U &= (\square,0)\\
V &= (\square,1)
\end{aligned}
$$

With these substitutions, the initial configuration in Figure 8-15 appears on a 1-track tape as $YWXY$. The following 1-track TM simulates the sum on a 2-track tape. (Compare to Figure 8–10.)

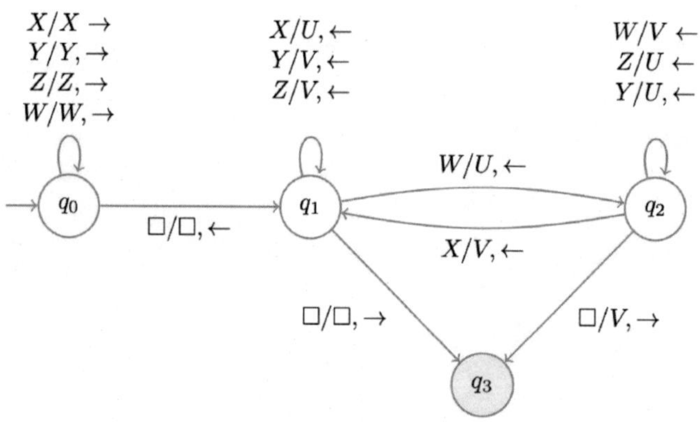

Figure 8–16: Simulating 2-track binary addition with a single track

The trace proceeds as follows:

$$
\begin{aligned}
(\mathrm{q}_0)YWXY &\vdash Y(\mathrm{q}_0)WXY \vdash YW(\mathrm{q}_0)XY \vdash YWX(\mathrm{q}_0)Y \vdash YWXY(\mathrm{q}_0)\Box \\
&\vdash YWX(\mathrm{q}_1)Y \vdash YW(\mathrm{q}_1)XV \vdash Y(\mathrm{q}_1)WUV \vdash (\mathrm{q}_2)YUUV \\
&\vdash (\mathrm{q}_2)\Box UUUV \vdash V(\mathrm{q}_3)UUUV
\end{aligned}
$$

The final ID above matches Figure 8–15, as expected.

The strategy for simulating an n-track tape for any n is the same—we invent a new symbol/variable for each possible combination of symbols in corresponding column positions in each track. The number of variables in the simulation may become large, but it will be finite.

A Turing machine with n tracks can be simulated with a standard, 1-track Turing machine.

A standard Turing machine can also simulate a TM with *multiple tapes*. The difference between multiple tapes and multiple tracks is that the read/write head on each tape can be positioned independently. The diagram below depicts how a multi-track TM can simulate a multi-tape TM. We will simulate three tapes: two for binary numbers to add, and one to write the result.

...		#	1	0	0	#	...
...		*					...
...	#	1	1	0	1	#	...
...	*						...
...					#	#	...
...					*		...

Figure 8–17: Simulating 3 tapes with 6 tracks

Each tape is represented by a pair of tracks, where one of the two tracks in a pair has blanks everywhere except for an asterisk associated with the cell where the read/write head points on its companion track above it. Since we can leave the asterisk in place except in one of the tracking tapes, we say that this simulation allows a *stay option*—we don't always have to move the read/write head on every tape on every move. This allows one tape to change position while others remain stationary, enabling independent movement among the tapes. It is convenient to delimit the used portion of the simulated tapes (we used a # above).

Now consider how to simulate the logic of the original 3-tape machine. We need to know the current symbol (above the asterisk) for each tape, so we position the read/write head for the simulating machine at the left extreme and make a left-to-right pass to determine what the symbol in track 1 is above the asterisk in track 2. We keep track of this by what state we move to in the simulating, multi-track machine. We repeat this process for the other pairs of tapes, ending in a state representing the unique combination of the three current symbols in each track. This tells us what each tape must do, so we make the appropriate changes in the tracks in a final pass. We repeat this process again and again until we reach a halt state. It sounds easy in concept, but the resulting machine is very complicated! The result is depicted in the following diagram.

...			#	1	0	0	#	...
...			*					...
...		#	1	1	0	1	#	...
...		*						...
...	#	1	0	0	0	1	#	...
...	*							...

Figure 8–18: Final tape contents

Since our new multi-track machine simulates a multi-tape machine, we can in turn simulate that multi-track machine with a standard TM, as explained in the beginning of this section.

We have used tape arguments to make these assertions, but you can imagine how to program them in any general-purpose programming language using an appropriate data structure (lists, arrays, maps, etc.). Since the Church-Turing Thesis asserts that languages that support structured programming are equivalent to Turing machines, it is reasonable to accept these variations of the standard Turing machine.

A Turing machine with n tapes can be simulated with a standard, 1-track Turing machine.

Finally, a standard, *deterministic* Turing machine can simulate a *non-deterministic* one. We explain this at the end of the chapter after we have examined the Universal Turing Machine, which models computers as we know them.

The Universal Turing Machine

The Turing machines we have seen so far are *single-purpose automata*—they only solve one problem or accept one language. Alan Turing went beyond that in his research to define a **universal machine**, a TM capable of simulating any other TM. The Universal Turing Machine (UTM) takes two inputs: a symbolic *encoding* of another machine, and an input string. It then simulates the execution of the input machine on the input string, and yields the result of that execution. This should sound familiar—that's exactly how general-purpose computers operate, and the UTM is where the idea of a stored-program computing mechanism originated. In other words, programs are just strings (rendered in some *encoding*) fed into a computer, and the computer runs the program on whatever input you give it.

> Just as a Turing machine models a specific computational process, the Universal Turing Machine models a computer.

Recall that the operation of a TM is defined by its transition function, δ. Each element in δ is a pairing of two inputs with three outputs, which we can represent as a 5-tuple. For example, the transition $\delta(q_0, a) = (q_1, b, \rightarrow)$ can be represented in text by the string "q_0, a, q_1, b, R".

Now consider how to write a program to act as a UTM using such input. Our program needs to know the start state, the halt state(s), and each transition in δ. The following text represents the input defining the TM in Figure 8–8, which rounds a unary number up to the nearest multiple of 3:

```
q0              # start state
q3              # halt state(s)
q0,1,q1,1,R     # these next 6 lines describe δ
q1,1,q2,1,R     # R = "right"
q2,1,q0,1,R
q0,B,q3,B,R     # B = "blank"
q1,B,q2,1,R
q2,B,q3,1,R
1111            # the input
```

The first two lines declare q_0 as the initial state and q_3 as the only halt state. The next six lines define δ. The last line is the input string to process. The output of the program is as follows.

```
(q0) 1 1 1 1
1 (q1) 1 1 1
1 1 (q2) 1 1
1 1 1 (q0) 1
1 1 1 1 (q1)
1 1 1 1 1 (q2)
1 1 1 1 1 1 (q3)
```

The logic for the UTM follows the pseudocode below:

UTM Pseudocode

```
read start state
read halt states
read rules
read initial tape    # the input string; will expand as needed
state = start
pos = 0              # start of string on tape
repeat forever:
    print id         # print the instantaneous description
    if state in halts
        break
    key = (state,tape[pos])      # 2-tuple
    rule = rules[key]            # rule is a 3-tuple
    state,tape[pos],dir = rule   # update the UTM
    calculate new pos based on dir
```

If we accept the Church-Turing Thesis, then we should be convinced at this point that such a UTM exists, having designed a structured program for it. Nonetheless, we will discuss conceptually how to construct an actual Turing machine that acts as a UTM. First, we must decide how to encode TMs as strings. There are many ways to encode a Turing machine. We will give a common one, using the tape alphabet $\Gamma = \{0, 1\}$.

The idea is to arrange elements of a Turing machine and its alphabet in sequences and use their respective sequence index numbers to represent them on the tapes of a 3-tape UTM. Look again at the TM in Figure 8–8 that we referenced above. There are four states, q_0 through q_3. We will use the numbers 0–3 to encode them. However, since we use *unary* numbers, we follow an *off-by-one* correspondence: $0 \to 1, 1 \to 11, 2 \to 111$, etc. We do this because we will use zeroes as separator symbols between tape elements. So, our states use the following encoding:

$$
\begin{aligned}
q_0 &\to 1 \\
q_1 &\to 11 \\
q_2 &\to 111 \\
q_3 &\to 1111
\end{aligned}
$$

For any input alphabet, $\Sigma = \{c_1, c_2, \cdots, c_k\}$, we use the corresponding unary indices $11, 111 \cdots (k+1)$-ones. In the current case, we have $\Sigma = \{1\}$, so we use the encoding 11. For a blank we use 1. For the directional indicators, we will use 1 for `L` ($\leftarrow$) and 11 for `R` ($\rightarrow$). We will load all of these onto three tapes, as follows:

1. Tape 1 contains the rules of δ, where each part of the 5-tuple for a rule is separated by 0, and each rule is separated by 00. For the current example, the first rule, "$q_0, 1, q_1, 1, R$" encodes as `1011011011011`. All six rules together form the single string (a backslash, \, continues the line):

```
10110110110110011011011101101100111011010101101100\
010111101011001101011101101100111010111011011
```

2. Tape 2 is the working tape for the TM being simulated, which we initialize with `11011011011` representing the input string 1111.
3. Tape 3 contains the current state of the input TM, initialized to 1 for q_0, followed by the halt state(s) (q_3 here). We again use 0 as a separator: `101111`. The prefix of this string gets replaced with the index of the current state as the machine progresses, and it is compared to the state(s) that follow to determine when to halt.

With all of this in place, the UTM follows the procedure outlined in the pseudocode above, using separators to find components as needed.

A TM can be encoded as a string

The strings containing the encodings of all transitions in δ for the set of all Turing machines form a *regular language.* We have the following pattern for our TM encoding:

```
<transition-1>00<transition-2>00...00<transition-n>
```

Each transition matches the regular expression $1^{+}(01^{+})^{4}$. The following regular expression, therefore, characterizes the language of all possible TM encodings using our encoding scheme:

$$1^{+}(01^{+})^{4}((001^{+}(01)^{+})^{4})^{*}$$

Non-Deterministic TM = Deterministic TM

With the notion of encoding, we can show how to simulate a non-deterministic TM (NTM) with a deterministic one. A non-deterministic TM is one where there are multiple choices for combinations of current state and character:

$$\delta(q_i, c) = \{(q_{k_1}, c_{k_1}, D_{k_1}), (q_{k_2}, c_{k_2}, D_{k_2}), \cdots, (q_{k_n}, c_{k_n}, D_{k_n})\}$$

In other words, for NTMs, δ is a *relation.* To simulate a NTM deterministically, we employ a *breadth-first traversal* of the tree of instantaneous descriptions representing all possible choices of applicable rules. Consider what happens at each step of a NTM, depicted by the sample graph below.

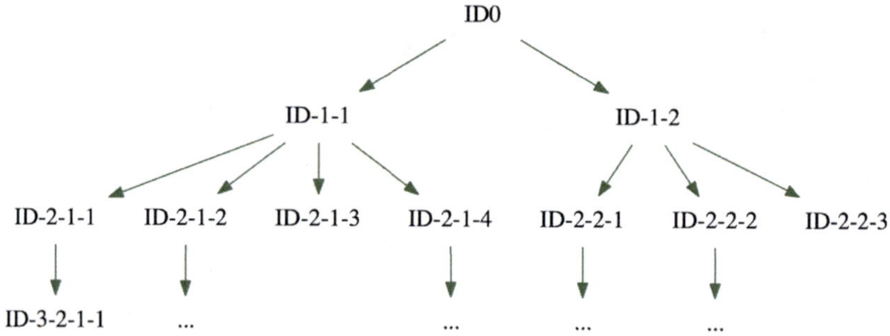

Figure 8–19: Tracing execution in a NTM

The way to follow the many choices in a NTM is to track the instantaneous descriptions that result at each step. We begin above with ID0, where the NTM is in the initial state and the read/write head is under the first input symbol. In the figure above, there are two choices at that point, leading to the configurations ID-1-1 and ID-1-2. These in turn lead to other possible configurations. We generate these IDs row-wise as we go and see if any configuration halts, such as ID-2-1-3, ID-2-2-3, and ID-3-2-1-1 above, in which case the simulation halts. If we ever reach a case where there is no choice to make, we crash. In this case a breadth-first traversal will halt at ID-2-1-3.

A breadth-first traversal implementation adds each node and its children to a queue, thus preserving a top-down, left-to-right ordering of the tree. Care must be taken to not visit a node twice. The following Python program implements a breadth-first traversal of the tree in Figure 8–19.

Python bread-first traversal of Figure 8–19

```
import collections

class Node:
    def __init__(self, name, children=[]):
        self.visited = False
        self.name = name
        self.children = children

def bfs(node):
    # Initialize queue with start node
    q = collections.deque()
    q.append(node)

    while q:
        node = q.popleft()
        if not node.visited:
            # Visit node
            node.visited = True
            print('visited', node.name)
```

```
        # Add this node's child nodes (if not visited yet)
        for c in node.children:
            q.append(c)

id0 = Node("ID0")
id11 = Node("ID-1-1")
id12 = Node("ID-1-2")
id211 = Node("ID-2-1-1")
id212 = Node("ID-2-1-2")
id213 = Node("ID-2-1-3")
id214 = Node("ID-2-1-4")
id221 = Node("ID-2-2-1")
id222 = Node("ID-2-2-2")
id223 = Node("ID-2-2-3")
id3211 = Node("ID-3-2-1-1")

id0.children = [id11, id12]
id11.children = [id211, id212, id213, id214]
id12.children = [id221, id222, id223]
id211.children = [id3211]

bfs(id0)
```

The output of this program verifies the expected ordering:

```
visited ID0
visited ID-1-1
visited ID-1-1
visited ID-1-2
visited ID-2-1-1
visited ID-2-1-2
visited ID-2-1-3
visited ID-2-1-4
visited ID-2-2-1
visited ID-2-2-2
visited ID-2-2-3
visited ID-3-2-1-1
```

The strategy for simulating the NTM is to use a tape to store the IDs in the order they are encountered by breadth-first search (in FIFO order, like a queue). Every time a choice occurs, the simulating machine copies the current ID onto another, working tape, and runs the original TM from there, encountering other IDs along the way, which are also added to the "queue" tape. When a halting configuration is found, the simulation halts, leaving the result on the working tape as part of the final ID.

A NTM can be simulated by a standard TM.

Key Terms

multi-track TMs • multi-tape TMs • universal TMs • encoding • breadth-first traversal • non-deterministic TMs

Programming Exercise

1. Implement the universal Turing machine as explained in this section. Use the input in the example to test your program.

Chapter Summary

A Turing machine is like the other types of automata we have studied, but it has a different memory model. The simple notion of being able to move to any location on a two-way tape of symbols enables TMs to perform any computation, and to recognize a larger class of languages than finite automata or pushdown automata. The transition functions of TMs constitute a very low-level but robust instruction set for computing.

Even more astonishing is the universal Turing machine, which can simulate any other TM. A UTM encodes TMs as strings, and embodies the conceptual basis for the development of the digital computer as we know it.

The remainder of this book will explore the consequences of Turing machines on our understanding of formal languages and on what is computable and what is not.

9. The Landscape of Formal Languages

Language is a process of free creation; its laws and principles are fixed, but the manner in which the principles of generation are used is free and infinitely varied. Even the interpretation and use of words involves a process of free creation.

– Noam Chomsky

Rocca Pietore, Italy, by Mauricio Artieda (pexels.com)

Where Are We?

Language Class	Recognized By	Generated By
Regular	Finite Automata	Regular Grammars Regular Expressions
Context-Free	Pushdown Automata	Context-Free Grammars
Recursively Enumerable	Turing Machines	**Unrestricted Grammars**

Chapter Objectives

- *Demonstrate the properties of Recursively Enumerable Languages*
- *Show the existence of languages that are not Recursively Enumerable*
- *Use Unrestricted Grammars to generate Recursively Enumerable Languages*
- *Understand the connection between Turing machines and Unrestricted Grammars*
- *Grasp the "big picture" of Formal Languages (the Chomsky Hierarchy)*

In this chapter, we complete our coverage of formal languages and their associated machines. The most general class of languages processed by automata are the Recursively Enumerable (RE) languages—those that are *recognized* by a Turing machine. There are important subcategories within the RE languages that we also explore. Most languages of practical use are in the subset of RE languages known as the *recursive* languages. We also lift all restrictions on rewriting systems to obtain grammars that generate the recursively enumerable languages. Finally, we show that there are more languages that are **non-RE** ("off the charts", if you will) than are RE.

9.1 Recursively Enumerable Languages

We begin this chapter with a definition:

> **Definition 9.1**
>
> A **recursively enumerable** language is a language that has an associated Turing machine that *recognizes* all and only the strings in the language. It may "hang" (run forever) when given an input string which is not in the language.

The term "recognized" is crucial. A TM recognizes an input string if it halts in an accepting state when processing it. The set of all such strings constitutes the language of the TM. If an input string is not in the language, the TM may (rarely) halt and reject it, or it may (more likely) get stuck in an infinite loop ("hang"). In other words, "it knows one when it sees one", but can't always tell when a string is *not* in the language. This is a consequence of the fact that TMs do not always halt. All the languages that we have seen so far are recognized by TMs that always halt (i.e., they are *decided* by the TM). We must do a bit of work to contrive a language that is RE but not recursive.

> **Definition 9.2**
>
> A **recursive** language is a language that has an associated Turing machine that *decides* all input strings. The machine *always halts*, explicitly accepting all the strings in the language and rejecting those not in the language.

In other words, there is an *algorithm* that decides whether or not any given string is in a recursive language. We say that the TM "decides" the language, instead of just "recognizing" it.

> A **recursively enumerable language** has a TM that *recognizes* it (it may hang on strings outside the language). A **recursive language** has a TM that always halts, *deciding* each string.

Since a recursive language's TM recognizes each string in the language, recursive languages are recursively enumerable by default and form a proper subset of the RE languages.

A Non-Recursive, RE Language

To get a "feeling" of a RE-but-not-recursive language, suppose you just made a public post on social media. Now consider the language consisting of the *handle* (username, email, or some other unique identifying string) of all persons who will at some point respond to your post. Notice the *future tense* here. Given an arbitrary handle of some user of the same social medium, you may wait forever to see if they respond. You can only know about those who have responded already. This language of such handle strings is not "decidable" because of the way it is defined, just as the function `g(f,x)` in the "Halting" subsection of Chapter 8 does not halt on all input strings because of the particular problem it is trying to model.

Context-Sensitive Languages

Nested within the class of recursive languages are the **context-sensitive** languages. A context-sensitive language is one where the size of the TM's tape needed to decide the language is *bounded.* The languages $\{a^n b^n c^n\}$ and $\{n_a(w) = n_b(w) = n_c(w)\}$ are each context sensitive. In both cases, the amount of the tape used is equal to the size of the input string, plus cells for a blank at each end of the string. A Turing machine that only uses up to a *constant factor* times its tape input size is called a **Linear Bounded Automaton** (LBA). The following table summarizes what we have explained so far about recursively enumerable languages. Each language class is a superset of the one below it. We will explain the grammar types in the next section.

Table 9–1: Hierarchy of Recursively Enumerable Languages

Language Class	Grammar	Machine
Recursively Enumerable	Type 0 (Unrestricted)	TM Recognizer
Recursive	Type 0 (Unrestricted)	TM Decider
Context-Sensitive	Type 1 (Non-contracting)	LBA

Properties of Recursively Enumerable Languages

We now explore key attributes of recursively enumerable languages to complete our perspective of formal languages.

First, *recursive* languages are *closed under complement.* This follows directly from the fact that if a language, L, is recursive, then there is a machine, M, that *decides* it. To decide its complement, $\overline{L}$, therefore, all we need do is invert the output results, as illustrated in the machine M' below.

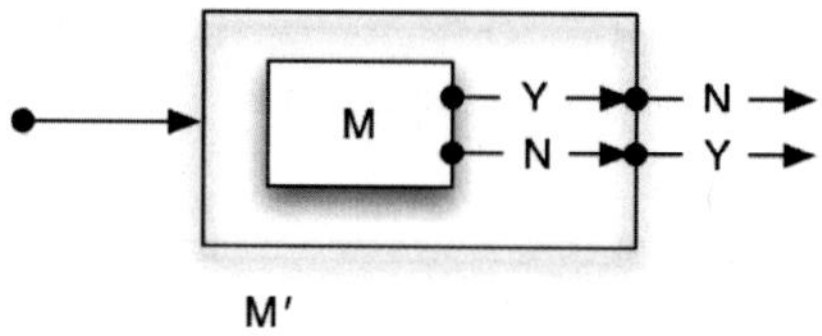

Figure 9–1: Deciding the complement of a recursive language

Recursive languages are closed under *complement.*

Now, suppose a language, L, and its complement, $\overline{L}$, are *both* recursively enumerable. This means that there are machines, M and M', say, that recognize each respective language. We can now form a new machine, M^*, that runs each machine *in parallel.* (See the diagram below.)

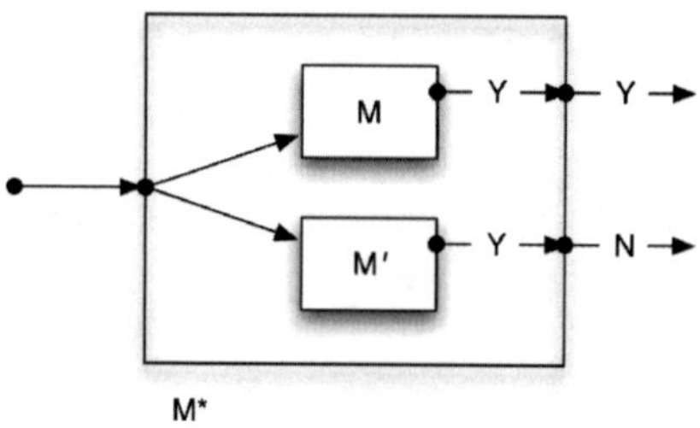

Figure 9–2: Running M and M' in parallel

By "running in parallel" we mean a cooperative multitasking scheme where each machine takes turns running a single instruction:

```
repeat:
    M executes its next instruction
    If M halted, halt and accept
    M' executes its next instruction
    If M' halted, halt and reject
```

Any input string is in one language or the other, so one of the machines is guaranteed to halt. If the string causes M to halt and accept, then M^* accepts it. If it causes M' to halt and accept, M^* rejects it. Therefore, the machine M^* *decides* L, which means that L and $\overline{L}$ are *both recursive*!

If a language and its complement are both **RE**, then both languages are *recursive.*

We can now conclude that the RE languages are not closed under complement. If they were, they would all be recursive languages!

RE languages are *not* closed under *complement.*

This means that the complement of a RE language that is not recursive is not even RE, a strange beast indeed!

Since RE languages are not closed under complement, they are also not closed under set difference, since the latter is defined in terms of the former.

Key Terms

recursive • recursively enumerable • context-sensitive • linear-bounded automaton

Exercises

1. Show that recursive languages are closed under union.
2. Show that recursive languages are closed under intersection.
3. Show that recursive languages are closed under concatenation.
4. Show that recursive languages are closed under reversal.
5. Show that RE languages are closed under union.
6. Show that RE languages are closed under intersection.
7. If L_1 is recursive and L_2 is RE, show that $L_2 - L_1$ is RE.

9.2 Unrestricted Grammars

What made context-free grammars "context-free" was that the left-hand side of each rule consisted of only a *single variable,* allowing substituting for that variable wherever it appears in a derivation, i.e., independent of its "context". An unrestricted grammar (aka "Type 0 grammar") allows multiple symbols, both terminals[1] and non-terminals, on the left-hand side of rules, allowing substitution to be sensitive to the context surrounding variables. For example, the rule $Ra \rightarrow aR$ allows *swapping* those two symbols in a derivation, but only applies when the variable R is followed immediately by the symbol a.

[1]Except for λ, which serves no purpose on a rule's left-hand side.

Definition 9.3

An *unrestricted grammar* is a rewriting system with rules of the form $s \to t$, where $s, t \in (\Sigma \cup V)^+$, and s contains at least one variable from V.

The following unrestricted grammar generates $\{a^n b^n c^n \mid n \geq 0\}$:

$$
\begin{aligned}
S &\to aXbc \mid \lambda \\
X &\to aXbY \mid \lambda \\
Yb &\to bY \\
Yc &\to cc
\end{aligned}
$$

Examine the following derivation to see how this grammar works.

$$S \Rightarrow aXbc \Rightarrow aaXbYbc \Rightarrow aabYbc \Rightarrow aabbYc \Rightarrow aabbcc$$

The variable S begins by generating one of each different character in the required order. The rule for the variable X is like the rule $X \to aXb$ for a CFG for $a^n b^n$, but it also appends a Y, which represents a c. However, the c's must go at the right end of the string, so the third rule above allows Y's to move to the right by swapping with b's until the Y bumps into a c, in which case the Y becomes a new c, keeping the counts of each letter in sync. Derivations using unrestricted grammars have a definite procedural feel.

Example 9–1

To generate the language of equal numbers of a's, b's, and c's, but in any order, the following grammar allows variables to be transposed arbitrarily.

$$
\begin{aligned}
S &\to XYZS \mid \lambda \\
XY &\to YX \\
YX &\to XY \\
XZ &\to ZX \\
ZX &\to XZ \\
YZ &\to ZY \\
ZY &\to YZ \\
X &\to a \\
Y &\to b \\
C &\to c
\end{aligned}
$$

Let's derive $ccabba$:

$$S \Rightarrow XYZS \Rightarrow XYZXYZS \Rightarrow XYZXYZ$$

Now that we have the correct number of proxy variables for the corresponding terminal symbols, we just move things around. First, we'll move the Z's to the left:

$$\Rightarrow XZYXYZ \Rightarrow ZXYXYZ \Rightarrow ZXYXZY \Rightarrow ZXYZXY \Rightarrow ZXZYXY \Rightarrow ZZXYXY$$

Now we swap the last two variables and finish off with terminals:

$$\Rightarrow ZZXYYX \Rightarrow^* ccabba$$

Example 9–2

The unrestricted grammar below[2] generates a^{n^2}:

[2]Adapted from Shallit, J., *A Second Course in Formal Languages and Automata Theory*, Cambridge University Press, 2009, pp. 174–175.

$$
\begin{aligned}
S &\to LGAR \\
L &\to LGAA \\
GA &\to aAG \\
Ga &\to aG \\
GR &\to R \\
L &\to X \\
XA &\to X \\
Xa &\to aX \\
XR &\to \lambda
\end{aligned}
$$

Observe the following about this grammar:

- The variables L and R stand for "left" and "right", and bracket the derivation
- The variable G generates an a every time it moves past an A on its right, but leaves the A in the derivation
- The variable L can turn into an X
- X "erases" any non-terminal to its right, passes by an a, and destroys itself and R when it hits R

This clever grammar uses the fact that each square number is the sum of a sequence of consecutive odd numbers:

$$
\begin{aligned}
1^2 &= 1 \\
2^2 &= 1+3 \\
3^2 &= 1+3+5 \\
n^2 &= \sum_{i=1}^{n}(2i-1)
\end{aligned}
$$

This grammar generates a single A to begin with, embedded in $LGAR$. Every time L is expanded using the second grammar rule, it generates two additional A's, guaranteeing that there will always be an odd number of A's in a sentential form. G's are placed strategically to achieve the summing of each successive odd number. Consider the derivation of $a^1 = a$:

$$S \Rightarrow LGAR \Rightarrow LaAGR \Rightarrow LaAR \Rightarrow XaAR \Rightarrow aXAR \Rightarrow aXAR \Rightarrow aXR \Rightarrow a$$

GA generates an a and the G moves past the A to the right. When G hits the right end, it vanishes. We then change L to an X and let it do its erasing, leaving a single a.

Let's now derive $a^{2^2} = a^{1+3} = a^4$, but this time we will keep track of the order in which the G's appear.

$$
\begin{aligned}
S &\Rightarrow LG_1AR \Rightarrow LG_2AAG_1AR \\
&\Rightarrow LG_2AAaAG_1R \Rightarrow LG_2AAaAR \\
&\Rightarrow LaAG_2AaAR \Rightarrow LaAaAG_2aAR \Rightarrow LaAaAaG_2AR \Rightarrow LaAaAaaG_2R \Rightarrow LaAaAaaAR \\
&\Rightarrow XaAaAaaAR \Rightarrow^* aaaa
\end{aligned}
$$

On the first line we have three A's, but notice where the G's appear. As G_1 moves to the right on line-2 it will generate one a, and G_2 will generate three a's on line-3, giving $1 + 3 = 4$ a's in total. If we were to derive $a^{3^2} = a^{1+3+5}$ we would begin as follows:

$$S \Rightarrow LG_1AR \Rightarrow LG_2AAG_1AR \Rightarrow LG_3AAG_2AAG_1AR$$

with G_1, G_2, G_3 generating one, three, and five a's respectively.

We now borrow an example from Rich[3] to further illustrate the power of unrestricted grammars.

Example 9–3

We will form a grammar to generate strings of the form $\{ww \mid w \in (a + b)^*\}$. The idea is to first generate wCw^R#, which can be done with the following *context-free* rules:

$$
\begin{aligned}
S &\rightarrow T\# \\
T &\rightarrow aTa \mid bTb \mid C
\end{aligned}
$$

Next, we will reverse w^R and place it after the hash symbol to obtain $wC\#w$, and then finally erase C and #.

To move the symbols of w^R to the right of the # backwards, we use a new variable, P, to "push" each character of w^R to the right, until P and its symbol reach the #, at which point it *inserts* the symbol immediately after the #, and then P disappears. The following additional rules accomplish the task.

---

[3]Rich, E., *Automata, Computability, and Complexity: Theory and Applications*, Prentice-Hall, 2008, pp. 513–515.

$$
\begin{aligned}
Ca &\to CPa \\
Cb &\to CPb \\
Paa &\to aPa \\
Pab &\to bPa \\
Pba &\to aPb \\
Pbb &\to bPb \\
Pa\# &\to \#a \\
Pb\# &\to \#b \\
C\# &\to \lambda
\end{aligned}
$$

The first two rules introduce the pusher, P, before the a or b to be pushed. Rules 3 through 6 move Pa or Pb to the right past an a or b. The last rule cleans up. We now derive $aabaab$:

$$
\begin{aligned}
S &\Rightarrow T\# \Rightarrow aTa\# \Rightarrow aaTaa\# \Rightarrow aabTbaa\$ \Rightarrow aabCbaa\# \\
&\Rightarrow aabCPbaa\# \Rightarrow aabCaPba\# \Rightarrow aabCaaPb\# \Rightarrow aabCaa\#b \\
&\Rightarrow aabCPaa\#b \Rightarrow aabCaPa\#b \Rightarrow aabCa\#ab \\
&\Rightarrow aabCPa\#ab \Rightarrow aabC\#aab \\
&\Rightarrow aabaab
\end{aligned}
$$

The first line generates $aabCbaa\#$. The next three lines push the characters of baa one at a time past the #, reversing baa to aab, and the last line cleans up the $C\#$.

The next example shows how unrestricted grammars can compute a *function.*

Example 9–4

If we allow an unrestricted grammar to begin derivations with an input string to process, instead of just expanding a start variable, we can simulate a *function.* The following grammar computes n div 2 for unary numbers.

$$
\begin{aligned}
S1 &\to T \\
T1 &\to 1S \\
TS &\to \lambda \\
SS &\to \lambda
\end{aligned}
$$

Like the TM in Example 8–4, this grammar erases every other 1. We begin by bracketing the input with S's, just like TMs bracket their input by blanks or other special symbols. We derive 5 div 2 and 6 div 2 below.

$$S11111S \Rightarrow T1111S \Rightarrow 1S111S \Rightarrow 1T11S \Rightarrow 11S1S \Rightarrow 11TS \Rightarrow 11$$

$$\begin{aligned} S111111T &\Rightarrow T11111S \Rightarrow 1S1111S \Rightarrow 1T111S \Rightarrow 11S11S \Rightarrow 11T1S \\ &\Rightarrow 111SS \Rightarrow 111 \end{aligned}$$

As you can see, S erases a 1 and T keeps it. By erasing the first 1 we automatically round down for odd numbers.

We call a grammar that computes a function a **computational grammar**.

Context-Sensitive Grammars

Recall that context-sensitive languages use a Linear Bounded Automaton, which is a TM that uses a region of its tape linearly proportional to its input size. The counterpart condition for grammars for context-sensitive languages requires that working derivation strings not grow excessively and then shrink back down to a terminal string. To guarantee that this does not happen, working derivation strings must be monotonically non-decreasing (i.e., they never decrease in length). To make good on this guarantee, lambda must only appear in the grammar to form the empty string, meaning that it only appears in the rule $S \rightarrow \lambda$, if it appears at all, and S must not appear on the right-hand side of any rule in that case, as we observed with Chomsky Normal Form. The other obvious requirement is that the right-hand side of a rule must *never be shorter* than the left-hand side. Such a "non-contracting", unrestricted grammar is called a *context-sensitive* (or Type 1) grammar.

The grammar for $a^n b^n c^n$ seen earlier in this section is not context-sensitive because X is *nullable*. A context-sensitive grammar is easily found, however, by substituting $X \mid \lambda$ for X and dropping lambda in X's rule, like we did when removing lambda from context-free languages:

$$\begin{aligned} S &\rightarrow aXba \mid abc \mid \lambda \\ X &\rightarrow aXbY \mid abY \\ Yb &\rightarrow bY \\ Yc &\rightarrow cc \end{aligned}$$

In the first rule, we replaced the first original lambda with the smallest, valid, non-empty string, *abc*. In the second rule, we replaced lambda with the expression abY. The next string in the language, *aabbcc*, derives as follows:

$$S \Rightarrow aXbc \Rightarrow aabYbc \Rightarrow aabbYc \Rightarrow aabbcc$$

We conclude that $a^n b^n c^n$ is a context-sensitive language.

The grammar in Example 9–2 is not context-sensitive—working strings grow beyond a constant times the size of the initial input before shrinking down to the final output.

Equivalence of Unrestricted Grammars and Turing Machines

As mentioned earlier, the behavior of unrestricted grammars is procedural, and any procedure can be simulated by a Turing machine. Therefore, for every unrestricted grammar there is a TM that *recognizes* (but not necessarily decides) its language or performs it function. To create a TM for an unrestricted grammar, we can use a non-deterministic TM with four tapes/memory areas containing the following:

1. Tape 1 holds the string to accept or reject
2. Tape 2 contains the on-going, working derivation
3. Tape 3 holds all the left-hand sides of the grammar rules
4. Tape 4 holds all the right-hand sides of the grammar rules, in the same order that the left-hand sides appear on Tape 3

To begin a derivation, we place the start variable, or starting configuration if we are simulating a function, on Tape 2, and execute the following logic:

```
repeat
    non-deterministically choose a cell on Tape 2
    non-deterministically choose the parts of a rule from Tapes 3 and 4
    apply the appropriate changes to the derivation on Tape 2
until the contents on Tape 2 matches Tape 1
```

If we place an invalid string on Tape 1, the procedure does not halt, but it halts on every valid input. This shows that the language of an unrestricted grammar is recursively enumerable. Likewise, any computational grammar can be simulated by a TM by placing the input on Tape 1 before running the machine.

For every unrestricted grammar, there is a Turing machine that recognizes the same language or performs the same function.

Going the other direction, it is straightforward to construct a computational grammar for a Turing machine by rewriting the TM's transitions in grammatical form, using the *states as variables*. We will convert the TM in Example 8–2 (adding two unary numbers) to a computational grammar. We repeat the transitions below in functional form for reference. (As a reminder, q_3 is the halt state.)

$$\delta(q_0, 1) = (q_0, 1, \rightarrow)$$
$$\delta(q_0, 0) = (q_1, 1, \rightarrow)$$
$$\delta(q_1, 1) = (q_1, 1, \rightarrow)$$
$$\delta(q_1, \square) = (q_2, \square, \leftarrow)$$
$$\delta(q_2, 1) = (q_3, \square, \rightarrow)$$

Transitions that move right are easy to simulate as grammar rules: we just need to move the state cursor to the right, while overwriting the symbol to the right, as TMs do. The first, second, third, and fifth transitions above yield the following grammar rules, respectively:

$$q_0 1 \rightarrow 1 q_0$$
$$q_0 0 \rightarrow 1 q_1$$
$$q_1 1 \rightarrow 1 q_1$$
$$q_2 1 \rightarrow \square q_3$$

When moving to the left, we need to consider that any possible tape symbol can *precede* the state cursor, so we have *multiple possibilities* for the fourth transition above:

$$0 q_1 \square \rightarrow q_2 0 \square$$
$$1 q_1 \square \rightarrow q_2 1 \square$$
$$\square q_1 \square = q_2 \square\square$$

$\square$ is a *variable* in the grammar that needs to eventually be removed, since it is not a valid language symbol ($\Sigma = \{1\}$ in this case). We therefore must add rules to remove blanks after reaching the halt state, as well as to remove the halt state after computing the numeric result:

$$q_3 \square \rightarrow q_3$$
$$\square q_3 \rightarrow q_3$$
$$q_3 \rightarrow \lambda$$

These ten rules comprise our computational grammar. We now derive the sum $3 + 4 = 7$ as we did in Example 8–2. We begin with the initial instantaneous description from that example, explicitly appending a blank as an end marker to mimic the TM.

$$\begin{aligned} q_0 11101111\square &\Rightarrow \to 1q_0 1101111\square \Rightarrow 11q_0 101111\square \Rightarrow 111q_0 01111\square \\ &\Rightarrow 1111q_1 1111\square \Rightarrow 11111q_1 111\square \Rightarrow 111111q_1 11\square \\ &\Rightarrow 1111111q_1 1\square \Rightarrow 11111111q_1\square \Rightarrow 1111111q_2 1\square \\ &\Rightarrow 1111111\square q_3\square \Rightarrow 1111111q_3\square \Rightarrow 1111111q_3 \\ &\Rightarrow 1111111 \end{aligned}$$

Other than erasing the blanks and the halt state, this grammar mimics the actions of the original TM.

Example 9–5

We now find a computational grammar for the TM in Example 8–3, which rounds a unary number up to the nearest multiple of 3. We again repeat the transitions below in functional form.

$$\begin{aligned} \delta(q_0, 1) &= (q_1, 1, \to) \\ \delta(q_0, \square) &= (q_3, \square, \to) \\ \delta(q_1, 1) &= (q_2, 1, \to) \\ \delta(q_1, \square) &= (q_2, 1, \to) \\ \delta(q_2, \square) &= (q_3, 1, \to) \end{aligned}$$

Each rule moves to the right, so the translation to an unrestricted grammar is straightforward:

$$\begin{aligned} q_0 1 &\to 1q_1 \\ q_0\square &\to \square q_3 \\ q_1 1 &\to 1q_2 \\ q_1\square &\to 1q_2 \\ q_2 1 &\to 1q_0 \\ q_2\square &\to 1q_3 \end{aligned}$$

Finally, we add rules to delete the remaining blanks and the halt state:

$$\begin{aligned} q_3\square &\to q_3 \\ \square q_3 &\to q_3 \\ q_3 &\to \lambda \end{aligned}$$

Observe the behavior of this TM with an input that is congruent to $1 \mod 3$, as in the trace of processing the number 4 below.

$$(q_0)1111 \vdash 1(q_1)111 \vdash 11(q_2)11 \vdash 111(q_0)1$$
$$\vdash 1111(q_1)\square \vdash 11111(q_2)\square \vdash 111111(q_3)$$

The first two moves on the second line above each process a $\square$, so we will delimit our input with two blanks at the right end when we begin a derivation. Here is the derivation for the number 4:

$$q_0 1111\square\square \Rightarrow 1q_1 111\square\square \Rightarrow 11q_2 11\square\square \Rightarrow 111q_0 1\square\square$$
$$\Rightarrow 1111q_1\square\square \Rightarrow 11111q_2\square \Rightarrow 111111q_3 \Rightarrow 111111$$

Example 9–6

We now construct a grammar for the TM in Figure 8–16, which adds two binary numbers. Since most of the moves are to the left, there will be a proliferation of rules to accommodate all the tape symbols that can precede the state cursor in an instantaneous description. We list each original TM transition followed by its associated grammar rule(s). First the right-moving transitions:

$\underline{\delta(q_0, X) = (q_0, X, \rightarrow):}$

$$q_0 X \rightarrow X q_0$$

$\underline{\delta(q_0, Y) = (q_0, Y, \rightarrow):}$

$$q_0 Y \rightarrow Y q_0$$

$\underline{\delta(q_0, Z) = (q_0, Z, \rightarrow):}$

$$q_0 Z \rightarrow Z q_0$$

$\underline{\delta(q_0, W) = (q_0, W, \rightarrow):}$

$$q_0 W \rightarrow W q_0$$

$\underline{\delta(q_1, \square) = (q_3, \square, \rightarrow):}$

$$q_1\square \rightarrow \square q_3$$

$\underline{\delta(q_2, \square) = (q_3, V, \rightarrow):}$

$$q_2\square \rightarrow V q_3$$

Now the left-moving transitions:

$$\underline{\delta(q_0, \Box) = (q_1, \Box, \leftarrow)}:$$

$$Xq_0\Box \to q_1X$$

$$Yq_0\Box \to q_1Y$$

$$Zq_0\Box \to q_1Z$$

$$Wq_0\Box \to q_1W$$

$$\underline{\delta(q_1, X) = (q_1, U, \leftarrow)}:$$

$$Xq_1X \to q_1XU$$

$$Yq_1X \to q_1YU$$

$$Zq_1X \to q_1ZU$$

$$Wq_1X \to q_1WU$$

$$\Box q_1X \to q_1\Box U$$

$$\underline{\delta(q_1, Y) = (q_1, V, \leftarrow)}:$$

$$Xq_1Y \to q_1XV$$

$$Yq_1Y \to q_1YV$$

$$Zq_1Y \to q_1ZV$$

$$Wq_1Y \to q_1WV$$

$$\Box q_1Y \to q_1\Box V$$

$$\underline{\delta(q_1, Z) = (q_1, V, \leftarrow)}:$$

$$Xq_1Z \to q_1XV$$

$$Yq_1Z \to q_1YV$$

$$Zq_1Z \to q_1ZV$$

$$Wq_1Z \to q_1WV$$

$$\Box q_1Z \to q_1\Box V$$

$$\underline{\delta(q_1, W) = (q_2, U, \leftarrow)}:$$

$$Xq_1W \to q_2XU$$

$$Yq_1W \to q_2YU$$

$$Zq_1W \to q_2ZU$$

$$Wq_1W \to q_2WU$$

$$\Box q_1W \to q_2\Box U$$

$\underline{\delta(q_2, X) = (q_1, V, \leftarrow):}$

$$Xq_2X \rightarrow q_1XV$$
$$Yq_2X \rightarrow q_1YV$$
$$Zq_2X \rightarrow q_1ZV$$
$$Wq_2X \rightarrow q_1WV$$
$$\Box q_2X \rightarrow q_1\Box V$$

$\underline{\delta(q_2, Y) = (q_2, U, \leftarrow):}$

$$Xq_2Y \rightarrow q_2XU$$
$$Yq_2Y \rightarrow q_2YU$$
$$Zq_2Y \rightarrow q_2ZU$$
$$Wq_2Y \rightarrow q_2WU$$
$$\Box q_2Z \rightarrow q_2\Box U$$

$\underline{\delta(q_2, Z) = (q_2, U, \leftarrow):}$

$$Xq_2Z \rightarrow q_2XU$$
$$Yq_2Z \rightarrow q_2YU$$
$$Zq_2Z \rightarrow q_2ZU$$
$$Wq_2Z \rightarrow q_2WU$$
$$\Box q_2Z \rightarrow q_2\Box U$$

$\underline{\delta(q_2, W) = (q_2, V, \leftarrow):}$

$$Xq_2W \rightarrow q_2XV$$
$$Yq_2W \rightarrow q_2YV$$
$$Zq_2W \rightarrow q_2ZV$$
$$Wq_2W \rightarrow q_2WV$$
$$\Box q_2W \rightarrow q_2\Box V$$

Since q_3 is the halt state, we add the final rules to remove q_3 and any remaining blanks:

$$\Box q_3 \rightarrow q_3$$
$$q_3\Box \rightarrow q_3$$
$$q_3 \rightarrow \lambda$$

With these 53 rules available, we begin with the same initial configuration immediately following Figure 8–16 to add $3 + 14 = 17$:

$$\begin{aligned}\Box q_0YWXY\Box &\Rightarrow \Box Yq_0WXY\Box \Rightarrow \Box YWq_0XY\Box \Rightarrow \Box YWXq_0Y\Box \Rightarrow \Box YWXYq_0\Box\\ &\Rightarrow \Box YWXq_1Y \Rightarrow \Box YWq_1XV \Rightarrow \Box Yq_1WUV \Rightarrow \Box q_2YUUV\\ &\Rightarrow q_2\Box UUUV \Rightarrow Vq_3UUUV \Rightarrow VUUUV\ (= 10001 = 17)\end{aligned}$$

This is the same output obtained from the original TM in Chapter 8.

Having a way to validate strings using a computational grammar for an arbitrary TM, we can feed into that grammar all possible strings in Σ^*, in effect generating all possible strings in the language of the original TM.

For every Turing machine, there is an unrestricted grammar that generates the same language or performs the same function.

Key Terms

Unrestricted (Type 0) grammar • context-sensitive (Type 1) grammar • computational grammar

Exercises

1. We said that the Context-free languages are a proper subset of the Context-sensitive languages. However, many Context-free grammars are contracting. How can any Context-free grammar be made non-contracting?
2. Create an unrestricted computational grammar that reverses any string of a's and b's. For example, starting with $SabbS$ you would end up with bba. (*Hint*: most of the work is done for you in Example 9–3).
3. Convert the following TM to an unrestricted, computational grammar.

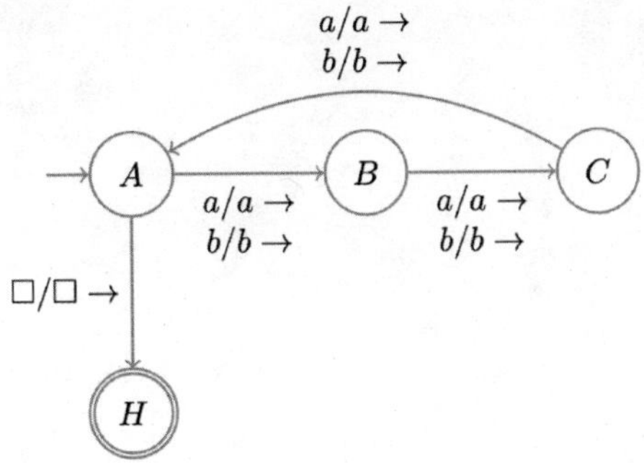

4. Using your grammar from #3 above, show the processing of the string $Aaba\Box$.
5. Using your grammar from #3 above, show the processing of the string $Aabba\Box$.
6. Convert the following TM to an unrestricted computational grammar.

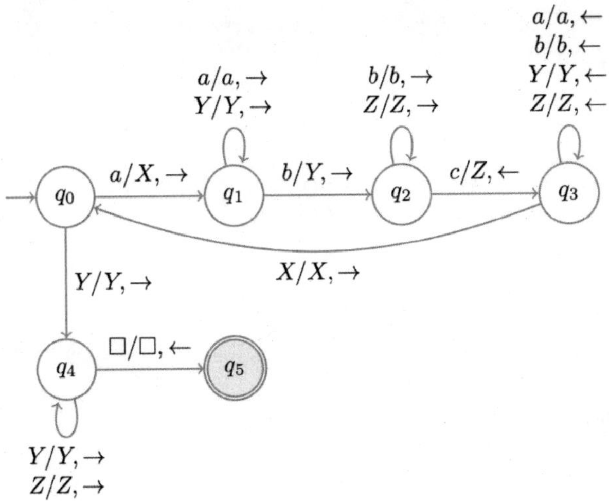

7. Using your grammar from #6 above, show the processing of the string $q_0aabbaa\Box$.
8. Using your grammar from #6 above, show the processing of the string $q_0abca\Box$.

9.3 The Chomsky Hierarchy

We can now summarize the entire *hierarchy* of formal languages. We emphasize "hierarchy" because each language class is a subset of the class above it. This classification originated with Noam Chomsky and is rightly called the **Chomsky Hierarchy**. See the following table.

Language Class	Recognizing Machine	Grammar Type
Recursively Enumerable	TM (recognizer)	Type 0: *Unrestricted*
Recursive	TM (decider)	
Context Sensitive	LBA (bounded TM)	Type 1: *Non-contracting*
Context Free	PDA	Type 2: *Context Free*
Deterministic Context Free	DPDA	
Regular	DFA	Type 3: *Right/Left Linear*

Table 9–2: The Chomsky Hierarchy of Formal Languages

We have yet to describe an RE language that is not recursive, and a language that is non-RE. That will still have to wait to the next chapter, but here we will show that there are more languages outside the Chomsky hierarchy (i.e., non-RE languages) than there are inside of it (the "recursively enumerable or better" languages). Our argument is based on the *countability* of infinite sets.

Countable Sets

A set is **countable** if there is a *one-to-one correspondence* between its elements and the *natural numbers*, $\mathcal{N}$ (or a *subset* of $\mathcal{N}$). Any finite set is therefore countable, and an infinite set is countable if its elements can be *enumerated*. The set of strings $\{a, b\}^*$ is countably infinite because its elements can be arranged in a well-defined sequence (i.e., the familiar *canonical order*):

$$e_0 = \lambda, e_1 = a, e_2 = b, e_3 = aa, e_4 = ab, e_5 = ba, \cdots$$

In this case, the subscripts are the natural numbers themselves, exhibiting the one-to-one correspondence between $\mathcal{N}$ and $\{a, b\}^*$ (i.e., $e(0) = \lambda, e(1) = a, e(2) = b, e(3) = aa, \cdots$).

We can conclude that all formal languages are countable sets, since they can be arranged in canonical order. A formal language is, after all, a *subset* of the countable set, Σ^*.

All formal languages are countable sets of strings.

Now, recall that Turing machines can be represented as strings over some alphabet. In particular, we showed in Chapter 8 how to encode any TM as a string over $\Sigma = \{0, 1\}$. Since sets of strings over a finite alphabet are countable sets, we see that the totality of possible TMs is a countable set.

The set of all possible TMs is countable.

We can now show that there are more languages (i.e., sets of strings over an alphabet) than there are Turing machines.

Uncountable Sets

An infinite set is **uncountable** if it is *not* countable. Since it is impossible to prove a negative directly, we typically show that a set is uncountable by assuming otherwise and then obtaining a contradiction. For example, the set of real numbers is uncountable. In fact, we can show that the set of real numbers in the open interval $(0, 1)$ is uncountable.

If the set of real numbers in $(0, 1)$ were countable, then we could *enumerate* its elements. The following diagram represents a hypothetical enumeration of those numbers, x_i, by arranging the digits of their (potentially infinite) decimal expansions in tabular form:

$$\begin{aligned} x_0 &= .d_{00}d_{01}d_{02}\cdots \\ x_1 &= .d_{10}d_{11}d_{12}\cdots \\ x_2 &= .d_{20}d_{21}d_{22}\cdots \\ x_3 &= .d_{30}d_{31}d_{32}\cdots \\ \ldots \end{aligned}$$

If the numbers in $(0, 1)$ were countable, then every single one of them would appear in this infinite list. We will now form a number in $(0, 1)$ that is *not* in the list to obtain the needed contradiction. Consider the number, x, with the following decimal expansion:

$$x = .((d_{00} + 1) \bmod 10)\ ((d_{11} + 1) \bmod 10)\ ((d_{22} + 1) \bmod 10)\cdots$$

We have added 1 to each diagonal digit in the table above, reducing it modulo 10 as needed, so the ith digit of x is $(x_{ii} + 1) \bmod 10$. x is a well-defined number in $(0, 1)$, but it differs from x_0 in its 0th digit, and from x_1 in its 1st digit, x_2 in its 2nd digit—in general, x differs from x_n in its nth digit for any n, so x, though a valid real number in $(0, 1)$, does *not appear* in the supposed enumeration. This *contradicts* the assumption that the numbers in that interval were enumerable, so we conclude that they are uncountable.

Next, we show that the powerset, 2^S, of any *infinite* set, S, is *uncountable.* It will suffice to show that the powerset of the natural numbers, $2^{\mathcal{N}}$, is uncountable.

Once again, if $2^{\mathcal{N}}$ were countable, we could enumerate its elements as a sequence of sets: $2^{\mathcal{N}} = \{p_0, p_1, p_2, \cdots\}$. We now form the following set of natural numbers, T:

$$T = \{i \mid i \notin p_i\}$$

The set T contains all the natural numbers, i, that do not appear in p_i. Since T itself is a set of natural numbers, it must appear somewhere in the enumeration, that is, $T = p_k$ for some k. We now ask the question, "Is k an element of p_k"? Well, if $k \in p_k$, then by the definition of $T = p_k$, $k \notin p_k$. Did you catch that? We have $k \in p_k \to k \notin p_k$. That is a *contradiction* if there ever was one! We likewise have $k \notin p_k \to k \in p_k$. We conclude that $2^{\mathcal{N}}$ is uncountable. The logic of this argument is the same no matter what infinite set we choose in place of $\mathcal{N}$.

The *powerset* of an infinite set is *uncountable.*

Now consider that the set of all languages over some alphabet, Σ. By definition, a formal language is a subset of the infinite set Σ^*. How many subsets of Σ^* are there? That number is $|2^{\Sigma^*}|$, the number of sets in the powerset of the infinite set Σ^*, which was just shown to be uncountable.

The number of languages over any finite alphabet is uncountable. Therefore, *there are more languages than Turing machines.*

Thus, there are more languages that are *not* associated with a TM than are, just like there are more real numbers than there are natural numbers. This proves the existence of (a whole lot of) non-RE languages.

Key Terms

Chomsky hierachy • countability • enumerable • uncountability • non-RE

Chapter Summary

The class of recursively enumerable languages encompasses all formal languages that we have studied. There are subsets with specific properties, but all useful languages fall somewhere in the Chomsky Hierarchy. All the examples we have seen so far in this book have been recursive or "better" (more specific/restricted). We have also seen that unrestricted grammars are equivalent to Turing machines in computational power. Finally, we showed that there are more non-RE languages than there are RE languages. After a long wait, we will see non-recursive and non-RE languages in the next chapter.

10. Computability

The question of whether computers can think is just like the question of whether submarines can swim.

– Edsger W. Dijkstra

Chapter Objectives

- *Exhibit a non-recursive language*
- *Exhibit a non-RE language*
- *Show that some problems/languages are undecidable/non-recursive based on the knowledge that other problems/languages are undecidable/non-recursive*

To complete our coverage of formal languages, we need to exhibit (finally!) a recursively enumerable (RE) language that is not recursive. We will form a language that can be recognized by a Turing machine, but that is not decidable (i.e., the TM may not halt on strings not in the language). This language will therefore be RE but not recursive. This language will allow us to discover other non-recursive languages, and, by complementation, languages that are not RE. But first, we consider some interesting features of self-reference.

> There once was a library intern tasked with creating a list of all books in the library that do not mention themselves. The intern found that the list became exceedingly long, and thought it best to expand the list itself into a book, to be housed in the same library for reference.
>
> Satisfied with her lengthy and tedious effort, she proudly placed the book of non-self-referencing books on a shelf in the reference section. After a while, it occurred to her and others that this new book itself, being a book in the library, also needed to be classified. Should this new book be listed inside itself? It certainly seemed so, because it did not

> mention itself within its pages. But once added, it now mentions itself, and therefore should be removed from itself, which mean it no longer mentions itself, so it should be added to itself...

This cute story is known as the *Librarian Paradox*. Can you see how it is reminiscent of what we did in defining the set $T = \{i \mid i \notin p_i\}$? in Section 9.3? Things get very interesting when *self-reference* and a *negative proposition* are involved. Here is another self-reference story of similar ilk:

> Here is the story of how I discovered that if you believe in psychic dogs, you must be barking.
>
> Ok, I've got a bunch of dogs that bark. Now these dogs aren't just any dogs, oh no, they're pretty clever. So, I've got a dog (Rex) that barks at cats, and a dog (Gnasher) that barks at fish, and a dog (Fido) that barks at black things, a dog (Pig) that barks at white things and some others. But am I happy? No. I want more!
>
> One day, I start playing with a mirror. Most of the dogs look in the mirror and aren't even remotely interested, but my dog that barks at black things (Fido) happens to be black, so when he looks in a mirror, he goes off on one. Some of the other dogs bark at themselves in the mirror as well.
>
> Unfortunately, my mirror is one of those full-length ones on the front of a wardrobe, so whenever I want to play with the mirror and the dogs I have to carry the whole wardrobe downstairs (the dogs aren't allowed upstairs).
>
> So, I think for some time, and I come up with a plan. I spend months training an amazing psychic dog (Meg), which barks if it sees a dog that will bark at itself. This makes it a bit easier, as I don't need to cart the wardrobe around anymore, Meg barks at Fido, but not at Gnasher, so I can take Meg around with me instead of going to all that furniture movement effort.
>
> One day, just for fun, I retrain Meg to bark at the opposite of what she used to bark at. This is fairly easy—she's pretty intelligent. Now, she barks at dogs that won't bark at themselves in a mirror. She barks at Gnasher now but not at Fido anymore.
>
> So, everything was fine, until one day, in contravention of the rules, Meg went upstairs and looked in the mirror...

10.1 The Halting Problem

The canonical example of a computation that is not decidable is the **Halting Problem**, which was originally proposed by Alan Turing himself. Suppose there is an algorithm, and therefore, a Turing machine, H, that takes an encoding of another TM, M, as input, along with an arbitrary string, w, and always halts, accepting if M will halt when processing w as input, and rejecting if it will not halt on w. This needs to be done without running M on w, of course, since that may hang, and then H wouldn't be an algorithm. H could be a very useful algorithm. We could call $H(M, w)$ to decide whether we should actually bother executing $M(w)$. This could lead to software that reads other programs and finds infinite loops, a nice, pre-emptive debugging tactic.

Intuitively, we suspect that the halting problem is unsolvable since TMs do not always halt. After all, what kind of *oracle*[1] can foresee whether a TM will halt on an arbitrary input string? But as usual, we would like a *proof.* We will show there can be no such H by assuming that it does exist, and then deriving a contradiction from that assumption.

If H exists, then we can use it as a subroutine to build other machines. For simplicity, we will write a new TM, H', that takes only one input, the string encoding of an arbitrary input TM, and feeds that encoding as both arguments to H. Put simply, $H'(M)$ calls $H(M, M)$. The input string is immaterial in what follows, so no harm done. The following figures depict H and H', respectively.

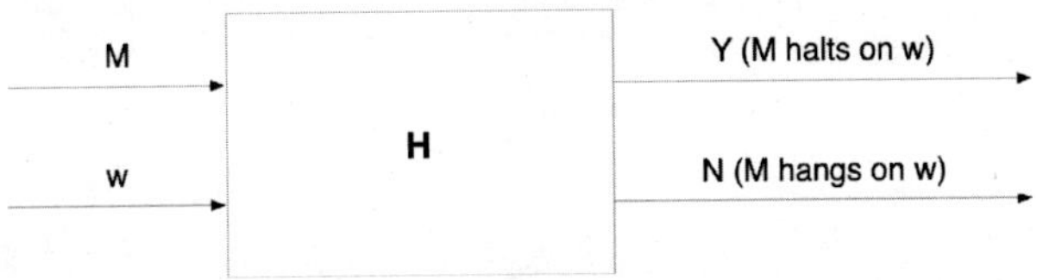

Figure 10–1: The machine H

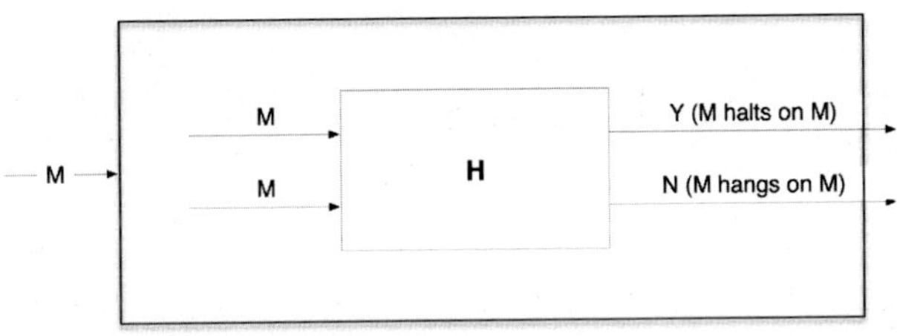

Figure 10–2: The machine H'

Finally, we create a new machine, D, which differs from H' in that where H' returns Y, D will instead go into an infinite loop. Otherwise it is the same. See the diagram below.

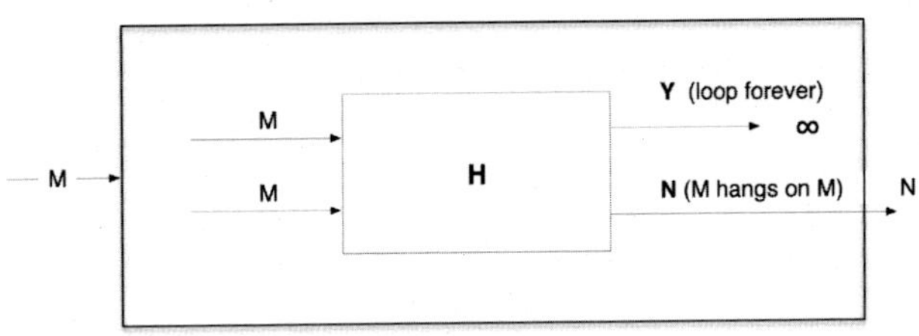

Figure 10–3: The machine D

A Python rendering of D looks like the following:

[1]An *oracle* provides answers to difficult questions without explaining how the answers were obtained.

The machine D in Python

```
def D(M):
    if H(M,M):
        while True:     # Infinite loop
            pass
    else:
        return 'N'
```

We now ask the question (here comes the self-reference...): what is the result of $D(D)$? Well, if D halts on its own encoding, then it loops forever (that is, it *doesn't halt*)! On the other hand, if it does not halt, then *it halts* with the output N. Any way you look at it, D is a logical impossibility. We are forced to conclude that H cannot exist!

The Halting Problem is undecidable

We can now define a recursively enumerable language that is not recursive. We first define a TM/program, R, that takes two arbitrary input strings, M, and w. If M is the encoding of a TM (that is, it matches the regular expression $1^{+}(01^{+})^{4}((001^{+}(01)^{+})^{4})^{*}$; review Section 8.3), then it runs M on w, otherwise, it returns N for "No". In the case that M halts on w, R returns Y, otherwise, it hangs. Here is a Python implementation of R:

The machine R

```
import re

def R(M,w):
    if re.match('1+(01+){4}((001+(01)+){4})*',M):
        M(w)     # May hang!
        return 'Y'
    else:
        return 'N'
```

We now define the language $L_H = \{\langle M, w\rangle \mid M \text{ is the encoding of a TM and } M \text{ halts on } w\}$[2]. This is the set of strings comprising the pairings of TM encodings and input strings where that TM halts on the given string. We know that this is a recursively enumerable language because R *recognizes* such combinations of M and w (i.e., it will halt on the subset of $\langle M, w\rangle$ where M halts on w). But L_H is *not recursive*, because R does not always halt; it hangs on strings that cause M to hang. An approach such as R is the only way to define this language because we know that the Halting Problem is unsolvable. L_H is recursively enumerable *but not recursive.*

[2]$\langle M, w\rangle$ denotes combining the encodings of M and w into single input string.

We can also state a related result: the language $L_d = \{\langle M \rangle \mid M \text{ rejects } \langle M \rangle\}$ is not recursive[3]. Suppose instead that some TM, M', say, decides L_d, and ask the question, "is $M' \in L_d$"? In other words, what does M' do when given $\langle M' \rangle$ as input? If M' accepts $\langle M' \rangle$, then it is, by definition, a member of L_d, because the string is accepted by M', the TM that decides L_d. But for $\langle M' \rangle$ to be in L_d means, by the definition of L_d, that it rejects $\langle M' \rangle$. Contradiction! Likewise, if it is not in L_d it is supposed to accept its encoding, which contradicts being accepted into L_d by M'. Contradiction again!

The language $L_d = \{\langle M \rangle \mid M \text{ rejects } \langle M \rangle\}$ is recursively enumerable, but not recursive.

We can now exhibit a non-RE language as well, namely, $\overline{L_H}$, the *complement* of L_H. If both L_H and its complement were RE, then they would both be recursive, as we showed in the previous chapter. But L_H is *not* recursive, due to the insolvability of the Halting Problem, so, $\overline{L_H}$ must be non-RE[4]. By the same logic, $\overline{L_d}$ is likewise non-RE, since L_d is RE but not recursive.

Key Terms

negative self-reference • undecidability • the Halting Problem

10.2 Reductions and Undecidability

Let's now consider the **Membership Problem**, which determines whether a given string is in a given language. For regular languages, we just run the string through the machine's DFA to see if it is accepted. For context-free languages, we can use the CYK algorithm to see if the string is generated by a given CFG. For recursively enumerable languages, we want an algorithm, A (for "accept"), that, given any TM M, and any string w, will determine whether M accepts w. Again, we doubt this is possible, since not all TMs halt. What would be the consequences of the existence of such an algorithm, A? In essence, it would mean that all recursively enumerable languages are recursive, since A would decide all TMs with an input arbitrary string. But we know that some recursively enumerable languages are not recursive (contradiction!), so A cannot exist[5].

The Membership Problem is undecidable.

[3] $\langle M \rangle$ denotes the string encoding of TM M.

[4] You may recall an earlier comment that non-recursive and non-RE languages were contrived entities. You were warned!

[5] This is a very strong result. This means that the membership problem *characterizes* RE languages. In other words, A is "reducible" to any other RE language's TM. Such languages are called *RE-complete*.

Knowing now that the membership problem is undecidable, suppose for the moment that we didn't know that the halting problem was undecidable. We can show that the halting problem, H, is undecidable by reducing the membership problem, A, to H. In other words, we can show that if H is decidable, then so is A—a *contradiction.* Consider the program (or, by the Church-Turing Thesis, the "Turing machine") below.

Reducing the Membership Problem (A) to the Halting Problem (H)

```
def R(M,w):
    if H(M,w):
        return M(w)
    else:
        return False
```

In this program, we assume that an algorithm for the halting problem, H, exists. If H indicates that M halts on w (i.e., it returns True), then R just executes M on w to determine whether it accepts w or not. On the other hand, if H indicates that M does not halt on w, then M can't accept w, so R rejects it. The algorithm R *decides the membership problem*, which we know is undecidable (contradiction!)—therefore, the halting problem must also be undecidable. We have *reduced* "solving" A to "solving" H. We use the notation $A \preccurlyeq H$ to mean that A is *reducible* to H, that is, solving A is no harder than solving H, because we have shown that if H is decidable, then we can use it to solve A.

In general, using the notation $\mathcal{D}(P)$ to mean that problem P is decidable, then

$$X \preccurlyeq Y \equiv \mathcal{D}(Y) \rightarrow \mathcal{D}(X)$$

This is what we typically do when solving problems algorithmically: we solve one problem in terms of another. For example, the factorial function, $F(n) = n!$, can be rendered in code as:

```
def F(n):
    return 1 if n <= 1 else n*F(n-1)
```

We have "reduced" solving $F(n)$ to being able to solve $F(n-1)$. Notationally, we say

$$\mathcal{D}(F_n) \preccurlyeq \mathcal{D}(F_{n-1})$$

Since we can certainly compute $F(n-1)$ (by completing the recursion down to 1), we can conclude that $F(n)$ is computable as well. In logical terms, we express this as[6]:

$$(\mathcal{D}(Y) \rightarrow \mathcal{D}(X)) \wedge \mathcal{D}(Y) \rightarrow \mathcal{D}(X)$$

If problem X is reducible to problem Y, And Y is *decidable*, then X is also *decidable.*

[6]This is the logical rule of inference known as "modus ponens".

What we did earlier with A and H was use the **contrapositive** of the reduction:

$$\mathcal{D}(Y) \to \mathcal{D}(X) \equiv \neg\mathcal{D}(X) \to \neg\mathcal{D}(Y)$$

In other words, we showed that $\neg\mathcal{D}(A) \to \neg\mathcal{D}(H)$. In general

$$(\neg\mathcal{D}(X) \to \neg\mathcal{D}(Y)) \wedge \neg\mathcal{D}(X) \to \neg\mathcal{D}(Y)$$

If problem X is reducible to problem Y, and X is *undecidable*, then Y is also *undecidable*.

To reiterate, if Y being decidable can be used to show that X is decidable, but we already know that X is undecidable, then Y must be undecidable[7]. This is how many problems relating to formal languages are shown to be undecidable. The challenge lies in coming up with a procedure to solve X using Y.

Example 10–1

Let E denote the problem of deciding whether a language accepted by an arbitrary Turing machine is empty or not—that is, E decides the question "$L(M) = \phi$?", where L is the language accepted by an arbitrary machine M. Note that an algorithm for E takes only a single Turing machine as input. We now show that the membership problem, A, is reducible to E by showing that we can solve A using E. The challenge is to convert the input to the membership problem, $\langle M, w\rangle$, to a single TM, M_w, say, so that $E(M_w)$ decides $A(M, w)$. Consider the reduction below.

Reduces A to E

```
def R(M,w):
    # We build a machine, Mw, whose language is the singleton {w}
    def Mw(M,s):
        accepted = M(s)      # True if and only if s ends in accepting state
        return accepted and s == w

    return not E(Mw)
```

The reduction program R takes M and w as input, and creates a new TM/function, M_w, that runs M on an arbitrary input string, s, and returns whether M accepted w (all other strings are ignored). In other words, M_w is a TM that runs M on any string, s, but can only accept w. Remember that M_w does not actually execute—the "oracle" E claims to "magically" return True if and only if M_w accepts *nothing.* (We don't care if $M(s)$ might hang—M doesn't execute either.) Having negated the output of E as we return from R, R returns True if and only if M accepts w (since w is the only thing M_w can accept). So R decides the membership problem, which is impossible. We conclude that E is undecidable.

[7]Here we are following the logical rule of inference, "modus tollens".

To summarize, the technique for reducing problem X to problem Y is to build an algorithm for X using Y, which entails transforming the input for X into input for Y in a way that a solution for Y leads to a solution for X. In other words, if x is input for X, then we seek to transform x into a string y such that $y \in L(Y)$ if and only if $x \in L(X)$. Like using the Pumping Theorems, this can be difficult and takes practice.

Example 10–2

The state-entry problem, S, asks if a TM ever passes through a given state, q, when given some input w. The halting problem reduces to this problem. We will create a new machine, M_q, that adds a new, unique state, q, to any TM, M, and has all the halt states of M move to state q, making q the new, unique halt state. Then we pass M_q to S.

Reduces H to S

```
def R(M,w,q):
    def Mq(s):
        # Build a machine that moves to new state q whenever M halts
        M(w)    # Run M on input w
        <code to move to state q>

    return S(Mq)
```

R takes three arguments: the TM encoding, an input string, and the state to check for entry. Since q is a new state not occurring in M, M_q halts in state q if and only if M halts on input w. So, if S can magically detect whether M_q would enter state q were it to process the input string w, then R solves the halting problem!

We conclude this chapter, and this book, by stating a result known as *Rice's Theorem*:

> **Rice's Theorem**
>
> Any **non-trivial** property of recursively enumerable languages is **undecidable**.

It is the fact that we have looked for algorithms—which by definition are TMs that always halt—to determine properties of computations that may not halt that has led to so many undecidable results. A "property" of languages equates to a *set* of languages (i.e., the languages that have that property). By "trivial", we mean properties that are either obviously always false for every RE language, or always true for every RE language. For example, we showed above that the problem, E, which determines if an arbitrary language has the property of "emptiness", is undecidable. We could have concluded this directly by Rice's Theorem, however, because some languages are empty and some are not. This is what is meant as a "non-trivial" property—one that holds for some TMs and not for others. All the reductions above involved non-trivial properties of TMs, and so are undecidable.

It is important to understand that Rice's Theorem only applies to properties of *languages, not machines.* Let the notation $n(M)$ denote the number of states in arbitrary TM, M. The following language/problem is recursive/decidable:

$$\{\langle M \rangle \mid n(M) \bmod 2 = 0\}$$

It is sufficient to inspect the encoding of a TM and count its states. The property "having an even number of states" applies to the *machine*—not to the language the machine accepts. On the other hand, the following *language*, which is the set of all TMs that only accept even-length strings, is not recursive:

$$\{\langle M \rangle \mid s \in L(M) \to |s| \bmod 2 = 0\}$$

Since some languages do contain only even-length strings (like $(aa + bb)^*$), and some contain odd-length strings ($(a + b)^*$), then Rice's Theorem guarantees that there is no algorithm that can decide the even-length property of all languages. It is sufficient to exhibit one language that has the property in question and one that does not.

Example 10–3

We can use Rice's Theorem to show that the problem of determining whether the language of an arbitrary TM is regular is undecidable. In other words, if R stands for the set of all regular languages, the following language is recursively enumerable, but not recursive:

$$\{\langle M \rangle \mid L(M) \in R\}$$

Since M can be any TM, and TMs recognize recursively enumerable languages, and regular languages are (vacuously) recursively enumerable, we can use Rice's Theorem. We must show that the property in question, "regular-ness", is non-trivial. But we know that there are regular languages, and there are also non-regular languages in the class of RE languages. Since "regular-ness" is true for some RE languages and not for others, this property is non-trivial and we conclude by Rice's Theorem that this problem is undecidable.

Key Terms

Membership Problem • reduction • non-trivial properties of RE languages • Rice's Theorem

Exercises

1. Show by reduction that determining whether an arbitrary TM/program accepts the empty string is undecidable. (*Hint*: use the Halting Problem.)
2. Show by reduction that determining whether an TM/program on input w ever reaches instruction n is undecidable. (*Hint*: consider Example 10–2.)
3. Show that the following language is not recursive.

$$\{\langle M\rangle \mid |L(M)| = \infty\}$$

4. Show that the following language is not recursive.

$$\{\langle M\rangle \mid |L(M)| \bmod 2 = 1\}$$

(In other words, $L(M)$ is finite and contains an odd number of strings.)

Chapter Summary

We showed that the Halting problem and the Membership problem are undecidable with proofs by contradiction. Many other problems can be shown undecidable by reducing a known undecidable problem to them. Reductions are the means of showing many problems to be undecidable, and thus are a very fruitful application of first-order, predicate logic. Provided that we can show that a property of interest for RE languages is non-trivial, we can use Rice's Theorem to conclude that there is no algorithm to decide that property over the class of RE languages.

Glossary

algorithm
: A TM/procedure that always halts, yielding output

automaton
: a finite-state machine

breadth-first traversal
: a graph traversal algorithm that visits all nodes at the same distance from a given node before preceding to the next level

Chomsky Hierarchy
: a classification of all types of formal languages

Chomsky Normal Form (CNF)
: a standard form for context-free grammars where each rule generates either a single terminal or exactly two non-terminals

Church-Turing Thesis
: the conjecture that any computation can be expressed as a Turing Machine

closure (of operators)
: the property that the result of any operations on elements from a set remains in that set

computable
: a function or language that can be processed by a Turing machine that always halts

computation
: a procedure that processes input and yields output (if it halts); the meaningful manipulation of symbols

computational grammar
: an unrestricted grammar that computes a function given an input string delimited by variables

context-free
: describes a language recognized by a pushdown automaton or generated by a context-free grammar (the latter has only a single variable on the left-hand sides of its rules)

context-sensitive language/grammar
: a language decided by a Linear Bounded Automaton, or generated by a non-contracting, unrestricted grammar

countable
: describes a set whose elements can be arranged in a one-to-one correspondence with (a subset of) the natural numbers

decidable
: a language or function for which there is a TM that always halts and accepts/computes the language/function

depth-first traversal
: a graph traversal algorithm that visits nodes in complete, left-most paths before preceding to other paths

derivation
: the process of generating a string from a grammar

determinism
: describes a process that operates the same way, without variation or ambiguity, every time the same input is given

DFA deterministic finite automaton; traverses the same path every time the same input is given

DPDA
: deterministic pushdown automaton

dynamic programming
: a procedure that works in stages, storing the results from each stage, so that subsequent stages may use results from previous stages

encoding
: a symbolic representation of set elements

formal language
: a set of strings from from symbols of a given finite alphabet

grammar
: a set of substitution rules for describing/generating strings in a language

halt state
: a state in an automaton that has no moves (out-edges); equivalently, it never appears in the domain of a transition function

instantaneous description (ID)
: of a Turing machine; a representation of the current state, the position of the read/write head, and the current contents of the used portion of the tape

lambda closure
: of a state in an NFA; the state itself along with the states reachable from that state by one or more lambda transitions

Kleene star
: zero or more concatenations of symbols of an alphabet or strings of a set

LBA Linear Bounded Automaton, a Turing machine that only needs a constant factor times the number of input symbols for working tape space

non-contracting grammar
: a formal grammar where the length of subsequent sentential forms in a derivation never decrease

non-deterministic finite automaton (NFA)
: a finite-state machine that allows zero or more transitions on a state for the same input symbol, as well as lambda transitions

non-terminal
: another name for a variable in a formal grammar; not a symbol in the alphabet in use

normalized PDA
a PDA where every lambda-pop transaction is accompanied by equivalent non-lambda-pop transitions popping each symbol of the stack alphabet

nullable
a variable in a formal grammar that can yield lambda in one or more substitutions

operator associativity
describes how adjacent operators of the same precedence are ordered (right vs. left associative, achieved with right or left recursion in a CFG)

operator precedence
describes the order of evaluation for adjacent operators (governed by where the associated rule appears in a CFG)

parse/derivation tree
a graphical representation of a derivation using a context-free grammar

partial function
a function defined only on a subset of its domain

post machine (queue machine)
a finite automaton that uses a queue for auxiliary storage

product table
a tabular representation of the simultaneous execution of two automata

pushdown automaton (PDA)
a finite automaton that uses a stack for auxiliary storage

recognizable
describes a recursively enumerable language or partial function that has an associated Turing machine that halts for only a proper subset of its domain

recursive language
a formal language decided by some Turing machine (always halts)

recursive variable
a variable in a formal grammar that appears both on the left and right-hand sides of the same rule

recursively enumerable language (RE)
a formal language that has an associated Turing machine that halts on strings in the language, but may not halt on strings not in the language

reduction
of one problem/computation to another. Used to show that a problem/computation is solvable/-computable or not in terms of another problem/computation

regular
describes a language represented by a regular expression or grammar, or that is decided by a finite automaton

regular intersection/union
the intersection/union of a context-free language with a regular language

subset construction
the process of tracking all possible output states together as a composite state when converting a NFA to a DFA

terminal
: a symbol in the alphabet of a formal language

total function
: a function defined on all input in its domain

transition
: an action in an automaton, represented by an edge in a transition graph, an entry in a transition table, or by a transition function

transition graph
: a directed graph that represents how states change when processing input symbols from an alphabet

transition table
: a tabular representation of an automaton, showing how input symbols map to states (and sometimes to auxiliary storage)

Turing machine
: a finite-state machine that uses a two-way, read/write, infinite tape for auxiliary storage

uncountable
: describes a set that is not countable

undecidable
: a language or computation for which there is no algorithm/TM to decide it (may hang on some inputs)

unit production
: a grammar rule of the form $V_1 \rightarrow V_2$, where V_1 and V_2 are single variables

unrestricted grammar
: a formal grammar that allows strings on the left-hand side of a rule instead of just single variables (enables contextual replacement)

useless variable
: a variable in a formal grammar that either is not reachable from the start variable, or that never yields a terminal string

Bibliography

1. Bernhardt, C., *Turing's Vision: The Birth of Computer Science*, MIT Press, 2016.
2. Brookshear, J., *Theory of Computation: Formal Languages, Automata, and Complexity*, Benjamin/Cummings, 1989.
3. Cohen, D., *Introduction to Computer Theory*, Second Edition, Wiley, 1997.
4. Enderton, H., *Computability Theory: An Introduction to Recursion Theory*, Academic Press, 2011.
5. Feynman, R., *Feynman Lectures on Computation*, Perseus, 1996.
6. Hopcroft, J., and Ullman, J., *Introduction to Automata Theory, Languages and Computation*, Addison-Wesley, 1979.
7. Kinber, E., and Smith, C., *Theory of Computing: A Gentle Introduction*, Prentice-Hall, 2001.
8. Lewis, H., and Papadimitriou, C., *Elements of the Theory of Computation*, Prentice-Hall, 1981.
9. Linz, P., *An Introduction to Formal Languages and Automata*, Sixth Edition, Jones & Bartlett, 2017.
10. Louden, K., *Compiler Construction: Principles and Practice*, PWS, 1997.
11. Martin, J., *Introduction to Languages and the Theory of Computation*, Fourth Edition, McGraw-Hill, 2011.
12. Minsky, M., *Computation: Finite and Infinite Machines*, Prentice-Hall, 1967.
13. Nagpal, C., *Formal Languages and Automata Theory*, Oxford University Press, 2011.
14. Parkes, A., *A Concise Introduction to Languages and Machines*, Springer-Verlag, 2008.
15. Rich, E., *Automata, Computability, and Complexity: Theory and Applications*, Prentice-Hall, 2008.
16. Scott, M., *Programming Language Pragmatics*, Fourth Edition, Morgan Kaufmann, 2016.
17. Shallit, J., *A Second Course in Formal Languages and Automata Theory*, Cambridge University Press, 2009.
18. Sipser, M., *Introduction to the Theory of Computation*, Third Edition, Cengage Learning, 2013.
19. Sudkamp, T., *Languages and Machines: An Introduction to the Theory of Computer Science*, Addison-Wesley, 1997.
20. Webber, A., *formal language: A Practical Introduction*, Franklin-Beedle, 2008.

Index

Made in the USA
Columbia, SC
10 September 2021

45241368R00168